29.95

DEDICATION

To our wives: Mary Anne Peterson and Dawn Ellen Beemer.

And, to our future:

Our children: Joy Kathryn Peterson Anckner, Nels Eric Peterson, Erik Linn Peterson, Eugene Ragnar Peterson, Lewis Charles Beemer, Nathan Eugene Beemer, and David Wendell Beemer.

Our grandchildren: Forest Zachary Franks, Jedediah Charles Franks, Ian Victor Franks, Jacob Noah Franks, Ellias Nels Peterson, August Graham Peterson, Seth Erik Peterson, and Victoria Marie Beemer.

First published in 1999 by MBI Publishing Company, 729 Prospect Avenue, PO Box 1, Osceola, WI 54020-0001 USA

© Chester Peterson, Jr. and Rod Beemer, 1999

All rights reserved. With the exception of quoting brief passages for the purposes of review no part of this publication may be reproduced without prior written permission from the Publisher.

The information in this book is true and complete to the best of our knowledge. All recommendations are made without any guarantee on the part of the author or Publisher, who also disclaim any liability incurred in connection with the use of this data or specific details.

We recognize that some words, model names and designations, for example, mentioned herein are the property of the trademark holder. We use them for identification purposes only. This is not an official publication.

MBI Publishing Company books are also available at discounts in bulk quantity for industrial or sales-promotional use. For details write to Special Sales Manager at Motorbooks International Wholesalers & Distributors, 729 Prospect Avenue, PO Box 1, Osceola WI, 54020 USA.

Library of Congress Cataloging-in-Publication Data

Peterson, Chester, Jr.
    American farm tractors in the 1960s / Chester Peterson, Jr. & Rod Beemer.
        p. cm.
    Includes index.
    ISBN 0-7603-0624-9 (hardbound : alk. paper)
    1. Farm tractors–United States–History. 2. Tractor Industry–United States–History I. Title.
TL233.6.F37P48 1999
629.225'2–dc21   99-38306

On the front cover: The International 806 was IH's answer to the call for increasing horsepower and capability. *Chester Peterson Jr.*

On the front flap: Ordered from Oliver in 1959 for a run of 500 tractors, and the same tractor as the Oliver 990, this 1960 98 diesel was one of the just 400 actually manufactured. Power is from a 371-cubic inch diesel engine and is 80 horsepower. *Chester Peterson Jr.*

On the frontispiece: The dash of a Allis-Chalmers D-21. *Chester Peterson Jr.*

On the title page: One of the best-known, not to mention cleanest-looking, members of the Oliver line, the 1900. *Chester Peterson Jr.*

On the back flap: The 1964 2000 offered 39 horsepower and a new paint scheme in a package that was essentially a refined version of the 9N introduced in 1939. *Chester Peterson Jr.*

On the back cover: Top: This 8010/8020 is one of only two mounting a cab, a damaged 7020 cab that Deere and Company converted to fit. This was the last 8010/8020 to leave the factory. Bottom: One of the stalwarts of the Allis line-up, the D-15. *Chester Peterson Jr.*

Edited by Lee Klancher
Designed by Tom Heffron

Printed in Hong Kong

CONTENTS

	ACKNOWLEDGMENTS	6
INTRODUCTION	NEW "ANTIQUE IRON"	7
CHAPTER 1	DEERE & COMPANY	13
CHAPTER 2	INTERNATIONAL HARVESTER COMPANY	41
CHAPTER 3	J. I. CASE COMPANY	57
CHAPTER 4	ALLIS-CHALMERS MANUFACTURING COMPANY	73
CHAPTER 5	MASSEY-FERGUSON, INC.	89
CHAPTER 6	FORD MOTOR COMPANY	105
CHAPTER 7	WHITE FARM EQUIPMENT COMPANY	121
CHAPTER 8	OTHER FARM TRACTOR COMPANIES	147
	BIBLIOGRAPHY	158
	INDEX	159

ACKNOWLEDGMENTS

Kicking off the decade with a "follow-me-if-you-can" attitude, the John Deere New Generation tractors of 1960 represented revolutionary rather than just evolutionary advances. Pacesetting the new lineup were the 3010 and 4010 models, destined to be some of the most popular tractors ever manufactured. These two both carry serial number 1000. This means they were the first 3010 and 4010 tractors to go out the factory door.

If you like the photographs of the many beautiful tractors of the 1960s, it's due to the efforts of these devoted restorers and craftsmen: Randy Addington; Larry Angtal; Ernie Burbrink; Kevin Burbrink; Roger Elwood; Dwight Emstrom; Wayne Findley; Leah Goecker; Randy Griffin; Michael Gross; J. R. and Jay Gyger; Darius Harms; Nathan Hill; Mike Hoffman; Edwin and Larry Karg; Walter and Bruce Keller; Jon Kinzenbaw; Paul and Nick Kleiber; Glen Knudson; Mel Kopf; Don Lamb; Ken Lang; Larry Maasdam; Paul Martin; McGinn Sales; Jeff McManus; Jerry Mez; Roger, Eugene, Martin, and Gaylen Mohr; Keith Oltrogge; Ken and Kent Peterman; Duane Peterson; Dennis Polk; Jack Purinton; Darold Sindt; Kenny Smith; Duane Starr; Ron Stauffer; Chris Tepoel; Brian Thompson; Roger and Gene Uhlenhake; and Doyle Whitney.

We also wish to express our thanks to the following people who were so generous with their time and knowledge: Eldon Brumbaugh, retired from J. I. Case Company after serving as director of corporate engineering and director of product performance and evaluation; Ralph Baumheckel, retired manager of International Harvester product planning research; Norm Swinford, retired from Deutz-Allis as product manager for tractors; James Ketelsen, president of J. I. Case Company 1967 to 1972; Del Gentner, retired from Massey-Ferguson after serving in product training, service training, and warranty; Ed Pinardi, retired engineer at Ford tractor division; Vincent P. Weber, retired as Senior Project Engineer at Oliver; Lee Vaughan, retired tractor project engineer at Minneapolis-Moline; Jw Richardson, production manager, M-R-S; Frank Howard, journeyman mechanic, Wagner Tractor Company; Ernest Nutsch, Allis-Chalmers collector; Don Horner, Ford collector; Bernard (Barney) Retterath, Oliver collector; Warren Wheeler, Cockshutt collector; Roger Uhlenhake, Oliver and Cockshutt collector; Guy Fay, International Harvester historian; Linda Davis, Salina Public Library; John Skarstad and Lee Phillips, University of California Davis special collections.

It has obviously required the cooperation of a goodly number of people to compile this work, and so we sincerely apologize to anyone we might have inadvertently not listed.

INTRODUCTION

NEW "ANTIQUE IRON"

Few icons are stirring up as much interest and nostalgia around the country right now as is the American farm tractor. Yet, until just recently, models chosen for a tractor "hall of fame" would undoubtedly have included only pre-1960s machines.

The reason for the cutoff was that tractors of the 1960s hadn't yet been tagged as "antique iron," perhaps due to their still modern styling and current work habits. A high percentage of these sturdy tractors are still being used every day on farms across the country.

Now, as then, however, perceptions are changing. The innovations and styling features that made 1960s tractors modern in their day are beginning to acquire the age and aura of the classic and to move this generation of tractors into the collector's realm.

The 1960s could justly be called a microcosm of the first 50 years of American farm tractor development. The changes in society and the marketplace, and the varied responses of the manufacturers to those changes, paralleled the evolution of the tractor throughout its history.

America's first successful tractor was probably the Charter tractor of 1889. "Successful" by this definition means that they were built, sold, and satisfied the farmer customer. The Froelich tractor, built in 1892, is also considered one of the initial successful American tractors. It was the first tractor to provide both a forward and a reverse gear; however, it failed the basic important criterion of satisfying the customer since the first two units sold were returned by unhappy owners.

Ten years later, many tractors were "successful" enough to entice scores of manufacturers to enter the tractor-building race as farmers began to purchase this new source of farm mechanization. In the next 20 years, as many as 200 manufacturers probed every imaginable combination of engine, transmission, and chassis configuration in pursuit of an affordable, practical tractor for American farmers. Most of these companies didn't succeed.

Some were underfinanced, some lacked engineering expertise, some lacked marketing and distribution skills, and some lacked business savvy. Where most faltered, though, was their inability, or refusal, to recognize and respond to present and future trends down on the farm.

All of these factors were complicated by social and economic conditions that ebbed and flowed across international borders. In America, more quickly than elsewhere, population growth and concentration were

By the mid-1960s, International Harvester had lost its number one status as a farm equipment manufacturer. The 56 series tractors showcased turbocharged engines, improved seat suspension, tilt steering wheel, and hydrostatic power steering in an effort to keep abreast of customer desires. This 1968 Farmall 856 is a 100-horsepower high-clearance example of the more powerful row-crop tractors in the International Harvester line.

Until 1954, Ford offered its customers only one model. When the company decided to expand the line, it did so a little too quickly. In 1961, the 6000 was rushed into production with disastrous results. The first 6000s were plagued with problems that required the company to replace the entire tractor except wheels and tires. This rare Ford 1967 6000 Commander is an LP gas–powered model.

away from rural farms and communities and toward urban centers. This left fewer farmers to produce more food and agricultural products with less available manpower.

World War I brought the situation into sharp focus. Not only were millions of men called into service, but also millions of horses, which were still a major means of moving military supplies to and on the battlefield. The shortage of manpower and draft

Allis-Chalmers was a vast, multi-faceted company that was occasionally stingy with funds for the farm equipment division. Innovations that were on the drawing boards in engineering sometimes languished years before they were incorporated into their tractors. The D21, however, managed to move Allis-Chalmers into the big tractor market with its direct-injection open-combustion-chamber diesel engine and an all-new powertrain. The new engine displaced 426 cubic inches and provided 103-PTO horsepower. This 1964 Allis-Chalmers D21 was the company's first tractor to offer more than 100 horsepower.

animals for farming curtailed agricultural production both at home and abroad.

Overseas, German U-boats sank thousands of tons of supplies headed for England. This created a severe food shortage that threatened to starve the United Kingdom into submission. As a result, Britain turned to American manufacturers for tractors in an effort to increase their own food production.

Before the conflict, some American tractors, such as International Harvester, Waterloo Boy, and others, were being exported to the United Kingdom; however, it was Henry Ford's Fordson that received the blessing of Britain's government. It was shipped across the Atlantic on a mass scale.

After the war, the market for tractors at home was brisk. Mechanized farming increased as the number of tractors on American farms rose from 10,000 in 1910 to 250,000 by the end of the 1920s.

Despite high demand, tractor manufacturers that had marginal financial resources couldn't compete and were forced out of business. Even those that did have substantial capital, such as General Motors, chose to cut their losses and discontinue manufacturing tractors. When the farm economy faltered in the mid-1920s, even more tractor manufacturers went out of business.

By the end of the decade, the number of remaining tractor manufacturers had fallen to roughly 30 fairly stable companies. International Harvester emerged as the front runner due largely to the row-crop Farmall's design. It gave the farmer a tractor that could do more than just replace the horse as draft power. Deere and Company's simple, dependable Model D kept them in the game. Case, Oliver, Minneapolis-Moline, Massey-Harris, Allis-Chalmers, and other lesser known names survived for the moment; however, serious social and economic pitfalls were just over the horizon.

The Great Depression, the drought and erosion of the "dirty thirties," and another world war again tested the mettle of these stubborn tractor manufacturers. Even through these trying times, farsighted industry leaders continued research and development programs that brought such improvements and refinements as power lifts, power take-offs, adjustable rear tread width, electric starters, and electric lights. Diesel engines appeared in the late 1920s and early 1930s and soared in popularity after World War II.

As the 1960s opened, Massey-Ferguson was number one in the world in wheeled tractor sales outside of North America. It held 23 percent of the Canadian market, but only 13.5 percent of U.S. sales. The company was led by Deere, International Harvester, and Ford. Its small to medium horsepower line was strong but the firm wasn't tooled to economically produce tractors in the high horsepower range. It had both Oliver and Minneapolis-Moline provide models in Massey-Ferguson sheet metal and colors. This 1961 95 Super was built for Massey-Ferguson by Minneapolis-Moline.

Case lost several points of market share in the 1950s and entered the 1960s with severe economic woes and a line of products, both agricultural and industrial, with design problems. It took a major amount of company time and money to correct the product lines. Representing the best that Case produced during the decade of the 1960s, these tractors stair-step in both size and their capacity to do fieldwork fast and efficiently. Pictured from left: 1964 830 Hi-Crop, 1967 1030 Wheatland Western Special, 1967 930 LP, 1963 730, 1962 Orchard and Grove, 1962 Orchard, 1965 530, 1962 430 LCK, 1964 Model 180 (garden tractor), and 1964 Model 130 (garden tractor).

Minneapolis-Moline was another company that opened the 1960s decade with major financial problems. Income had eroded by 50 percent from the boom years of the 1950s—down from $107 million in 1953 to $49 million in 1960. Coupled with the engineering emphasis placed on the unsuccessful Uni-Tractor, this resulted in Minneapolis-Moline's sale to the White Motor Corporation in 1963. This Minneapolis-Moline lineup includes two large Heritage models that display a different and distinctive paint scheme instead of the company's traditional yellow or Prairie Gold motif. Pictured from left: 1969 A4T-1600 Plainsman, 1969 White G-950, 1969 A4T-1600, 1969 G1050, 1969 G1000 Vista, 1964 G706, 1964 M670, 1964 M602 Propane, 1965 U302 Super, 1966 Jet Star 3 Super, 1961 M5, 1960 4 Star Super, and 1961 Motrac.

What tractors looked like received extremely little consideration before the 1930s. Form followed function—at quite a distance sometimes. It was obvious to most manufacturers that tractor appearance wasn't important just as long as the tractor got the crops planted and harvested expeditiously. This began to change as both International Harvester and Deere & Company made the decision to hire industrial designers to give their tractors more "eye appeal."

In 1939, the famous "handshake agreement" between Henry Ford and Harry Ferguson (which was never formalized in writing) ushered in a radical new concept in tractor and implement design. The resulting 9N Ford tractor with Ferguson's hydraulic system and three-point mounted implements literally revolutionized the industry.

Although some, such as Case's President Leon R. Clausen, considered the small Ford-Ferguson tractor and mounted implements just a gimmick, the concept was eventually copied and imitated by all tractor manufacturers. Some manufacturers, however, waited more than a decade to respond, losing important sales for their misreading of market trends.

Immediately after World War II, tractor manufacturers again enjoyed a bullish sellers' market. During the conflict, manufacturing facilities were shifted to war production, and tractor manufacture was drastically reduced. Europe's tractors were either direct victims of the war or victims of neglect and overwork. And in America, tractors that had not been replaced during the Depression or before the war were showing their age.

Postwar demand enticed some 22 new companies to enter the tractor manufacturing business, but demands had changed. Farm size was increasing, requiring bigger and more powerful tractors to maintain the larger operations. In the late 1940s, most of America's farmland was worked with tractors in the 20- to 30-drawbar horsepower range. At the close of the 1950s, the upper range for production tractors was in the neighborhood of 70 horsepower. Bucking the established full-line companies proved too great an obstacle to overcome, and by the late 1950s, all of the newcomers were gone. Regardless, the boom market had subsided by 1953.

A negative aspect of the robust postwar economy was that it lured some tractor manufacturers to focus on production and sales of existing models to the neglect of research and development. In some cases, embryonic tractor manufacturing was a stepchild of companies heavily capitalized in other pursuits apart from a full line of agricultural equipment. Expansion and diversification by these corporations in other areas channeled profits away from future tractor development at a time when it was really needed.

Falling into this same trap, some postwar tractor and implement companies seemed to invest in everything, from fertilizer plants to refrigerators to air-conditioners to home heating units to milking machines, cream separators, and aircraft engines. None of this furthered their tractor business.

During the early 1960s the spirit of change that began after World War II was at full throttle. Almost everything was being challenged during this turbulent decade—authority, morals, the military, politics, government. This was reflected in the Vietnam War, Kent State, the assassination of President Kennedy, the space program putting a man on the moon, and, yes, rapidly evolving farm tractors. Farm size continued to increase, bringing calls for still more powerful tractors. As the 1960s ended, a farmer could take home any number of 100-plus-horsepower tractors right off the dealer's showroom floor. A few limited-production tractors had even reached the brawny 300-horsepower level.

This trend was toward a tractor that was not only bigger, but also more advanced in every respect. More massive equipment demanded bigger and better hydraulic systems. Equipment driven by the power take-off called for more versatility from the PTO design too.

Larger engines burned more fuel and so more capacious fuel tanks had to be mounted. More powerful engines necessitated bigger and improved tire designs to harness the increased horsepower and improve traction while lessening the effect of compaction.

Farmers began demanding transmissions with a wider range of speeds to better suit a variety of working conditions. The traditional four-speed gearboxes just weren't acceptable anymore. Shift-on-the-go transmissions that could match the gear to the job without clutching became a must-have item.

Operator comfort and safety were major issues that also significantly impacted tractor design for the first time during the 1960s. Power steering and power brakes made handling easier and safer and improved seats reduced fatigue. Rollover Protective Structures (ROPS) were designed to protect operators from serious injury in a rollover.

Tractors, underneath whatever paint color, were converging on a similar basic design in the 1950s. Yet there was still plenty of room for innovations that could set one model apart from its competitors. Some companies weren't as quick to design or get to market with these innovations as others, though.

As Robert T. Kudrle observed in *Agricultural Tractors: A World Industry Study*, "The introduction of power-steering, higher horsepower, four-wheel-drive, and other rather obvious developments did indeed shift market shares. But, there is little in any of the innovations which any firm in the industry could not easily have been the first to introduce if it had chosen to do so."

These so-called innovations, coupled with dependable major components, became increasingly important as farmers placed increasing demands, and increasing reliance, on their machines. By the late 1960s, tractors accounted for 35 percent of all farm machinery and equipment expenditures in the United States. From the early 1950s to the close of the exciting 1960s, the players were the same. Mar-

Oliver's famous Fleetline series first appeared in 1949 and the smooth six-cylinder engines became hallmarks of the innovative Oliver tractors. By 1960, however, the company was floundering and White Motor Corporation purchased the tractor plant, tillage equipment division, and the harvesting equipment facility, plus other assets of the corporation. Under the new ownership arrangement, the Oliver name continued to appear on tractors until mid-1970s. This 1964 Oliver 1600 is fitted with a 1610 front-end loader.

ket share had shifted significantly, however. Ownership was often altered through mergers and sales, although the major companies managed to survive.

The 1960s represent an important chapter in the history of the tractor. These machines have more power and many modern features, but also represent the twilight years of several makes and models that made important contributions to American agriculture. This decade produced some fine tractors, which well deserve the growing attention and appreciation they now receive.

CHAPTER ONE

DEERE & COMPANY

John Deere didn't live to see his name emblazoned on the hood of a tractor. Nor did his son, Charles Deere. But that name endured to become one of the premier symbols of the farm tractor.

John Deere was born February 7, 1804, in Rutland, Vermont. He died May 17, 1886, in Moline, Illinois. His numerous trials, hardships, and decided triumphs in those 82 years have filled many a book and receive only brief treatment here.

Fatherless by the age of four, Deere learned early to look out for himself. He was an apprentice blacksmith in his youth and married in 1827. He opened several blacksmith shops as a young man. Two of these, both in Leicester, Vermont, burned down in quick succession. He found work elsewhere, but accumulated debts caught up with him, putting him on the verge of arrest. Deere then left his growing family, and debts, to pursue his dream on the frontier. At his blacksmith shop in Grand Detour, Illinois, opened with business partner Leonard Andrus, Deere created the famous John Deere steel plow.

In 1848, Deere relocated to Moline, Illinois, where the business was named Deere and Company in 1857. A year later, Deere deeded his full interest in Deere & Company to his son, Charles Deere, a move to keep the financially strapped company in business. Under Charles' leadership, the company thrived, becoming a full-line, full-service company by the turn of the century.

Deere developed its first tractor over the years 1912 to 1917, behind such competitors as International Harvester and Case. The very early "Dain" models, authorized for production in 1918, had two forward and two reverse speeds, driven by a 12-horsepower, four-cylinder gasoline engine. The venerable Model D—a refined version of a tractor produced by Waterloo Boy, an acquired competitor—followed in 1924 and stayed in the lineup until 1953. It was joined by a smaller, general-purpose model, the C, in 1927, and by the extremely successful Model A, two-plow tractor, in 1934. Other models followed in the 1930s to handle a variety of more specialized farm operations, making use of hydraulics and the power take-off.

The war years and immediate postwar period brought only minor variations to the existing model line, including added fuel options, electric starter, and electric lights. To compete with the popular Ford-Ferguson, Deere introduced the Model M in 1947, produced at its new plant in Dubuque, Iowa. The Model R, Deere's diesel-powered replacement for the D, followed in 1949. It was Deere's highest power model at the time, with 34 horsepower at the drawbar and 43 at the belt.

Deere & Company's first "numbered" tractors appeared in 1952 as the 50 and 60, replacing the A and B models. The 70 followed, along with the 40 and 40C in 1953, and the big 80 Diesel joined the lineup in 1955. The 20 Series debuted in 1956, featuring the 320, 420, 520, 620, 720, and 820. By this time refinements and upgrades of numerous features were available, such as live power shafts, LP gas, hi-crop units, a variety of front-end and rear-axle options, improved power steering, improved hydraulics, and added horsepower.

Operator comfort and convenience were the focus of the 30 series, which was in showrooms by 1958. The 30 model's new seat, new fenders, new

The 1960s was a John Deere decade. The company opened the time period by introducing a sophisticated line of four- and six-cylinder tractors called the New Generation. This 2010 Hi-Crop is one of the models in that line, a series of tractors that would put John Deere into the number one manufacturing slot. The company has remained an industry leader since that time.

13

Rare to begin with since only 350 were made, "Big Bertha," a 1960 830 Diesel Rice Special, was one of the last non-New Generation tractors Deere & Company ever produced. The revolutionary New Generation tractors were released in August 1960. This particular tractor came equipped with power steering.

Only a few of the early 1010 utility tractors, like this one built in November 1960, came equipped with optional power steering. The gearshift lever for the five-speed transmission is on the console. Its clutch would often stick if unused for a lengthy time—a tendency it shared with other tractors made at Deere's Dubuque, Iowa, plant. The slightly less than 50-inch shoulder height of the hood line was promoted in advertising.

lighting, new instrument panel, and tilted steering wheel previewed state-of-the-art "human factor" design that would follow shortly on the "New Generation" tractors.

The first New Generation model, the articulated four-wheel-drive 8010, hit the market in the last year of the 1950s and caused quite a stir. Showcased at a Deere & Company field day near Marshalltown, Iowa, the 200-plus horsepower monster stood more than 8 feet tall. It was 8 feet wide, 19 feet long, and weighed only 300 pounds short of 10 tons. "Too big," said some farmers, but they rushed to attend a field day demonstration, where the new machine pulled a bigger plow and threw more furrows than they thought possible. The company was forced to rent or lease most of these monsters, and had to refurbish the driveline on all but one of them, but the stage was set.

John Deere could never have foreseen where his company and his drive for success would lead. A new generation of powerful and comfortable tractors was at hand—a generation that would take shape with the machines of the 1960s.

THE 1960S AND DEERE'S NEW GENERATION

The New Generation tractors, introduced in 1960, put Deere & Company several years ahead of its competition. These new tractors carried industry firsts, such as high horsepower-to-weight ratio, hydraulic power brakes, closed-center hydraulics, and lower-link sensing on the three-point hitch system. Another big advancement was the introduction of hydrostatic power steering. This eliminated the traditional mechanical link between the steering wheel and front wheels.

Many of the engineers who were responsible for creating the New Generation tractors still live around the Waterloo, Iowa, area enjoying their accomplishments and retirement. Mike Mack, retired director of Deere & Company's Product Engineering Center (PEC), knew many of the key players in the New Generation drama.

"Merlin Hansen is generally considered to be the father of the New Generation tractors," said Mack. "He headed up the project when the PEC opened in 1956. Others who played key roles were Wally Dushane who managed design, Don Wielage headed

Although the smallest tractor in the revolutionary New Generation line, this 1961 1010 developed 35-PTO horsepower, and harnessed enough muscle to pull a three-bottom plow through most soils. The 1010 was manufactured at Deere & Company's facility in Dubuque, Iowa. A big percentage of the 1010 components were built on the old Model M tooling, and as a result didn't share some of the advanced features of the 3010 and 4010.

15

Every new owner of a New Generation John Deere tractor received a comprehensive "Operator's Manual." This particular manual is specifically aimed at the owner of a new 1010 single Row-Crop tractor, although it's generically similar in appearance to the manuals of the other models.

up development, Sid Olsen was responsible for engine design, Vern Rugen was in charge of transmissions, Danny Gleeson oversaw chassis and controls, Ed Fletcher headed hydraulics, and Chris Hess led the group working on the hitch."

What did these men, and CEO Bill Hewitt, consider true New Generation tractors? Interestingly, they felt that only the 3010 and 4010 truly represented the New Generation concept. True, the 1010 and 2010 were of the same family, but they didn't have all the revolutionary features of the 3010 and 4010 that were incorporated into the rest of the New Generation line, worldwide, at a later date.

Another surprise was their reaction to a question concerning what they thought was the greatest design advancement on the New Generation. Without hesitation the reply was, "The hydraulic system. If it wasn't the greatest, it had to be right up there at the top."

This new hydraulic pump and system allowed the New Generation tractors to leap-frog years ahead of the competition. At the time, conventional gear pumps were usually only able to maintain 1,000 pounds per square inch (psi) of pressure. Deere's new eight-piston design could deliver 20 gallons a minute at 2,000 psi. This allowed rock shaft and power steering cylinders to be smaller, while still delivering plenty of power for the increasing number of functions hydraulics were now being asked to perform. The transmission and differential case provided a reservoir for the system, using a specially designed oil that served as the hydraulic fluid and gear lubricant. A system filter and oil cooler were also part of the package.

Perhaps the biggest departure from the "standard" John Deere tractor was the new engines. The benchmark 3010 and 4010 tractors were both powered by the new Deere & Company–designed and –built engines.

Mack commented on the then-new four- and six-cylinder development as he pointed to a cylinder block on the line at Deere's Waterloo Engine Works in 1998, saying, "Sid Olsen really was the man behind all the engines that are still built today. They're all a derivative of what he designed back in the early 1950s. The engines have gone up in displacement—you stroke 'em and bore 'em—but the centerline distance from cylinder to cylinder is still what Sid laid out back then. He was a really, really good engine designer."

Roll-O-Matic dual front wheels are featured at the front end of this 1961 Row-Crop, and as yet only partially restored, 1010. Other front axle options were the adjustable utility straight axle and the swept-back utility adjustable axle. The swept-back axle provided a shorter wheelbase for easy maneuvering.

A new concept of tractor power

New low-budget surprise package 1010 delivers up to 35 horsepower
—if you need it

These are the "compacts" in the new John Deere New Generation of Power. And they are compacts with surprise features that make them big in earning power.

Most important is the new John Deere-built variable-speed engine that puts a new kind of power into your farming operation. Variable-speed means variable-horsepower —and in these tractors, you meter engine power exactly to the job at hand. You get high efficiency in field work at any engine speed from 1500 to 2500 rpm. In terms of power, that's 23 to 35 horsepower, measured at the power take-off.

The result: You'll pull a 3-bottom plow in most soils, at a fast clip, with the engine revved up. Yet, when you handle a light load, you can throttle down and slip the transmission into a high gear for fast work with maximum fuel economy. These surprising new 4-cylinder engines come as either Diesel or gasoline versions, in Single Row-Crop or Utility Tractors.

That's just a part of the "1010" surprise package. You'll vote approval of such features as comfortable, adjustable seats, power steering, and handy controls, including the gearshift lever conveniently mounted on the tractor dash.

You'll like the new independent "live" 540-1000 rpm power take-off, with its hand clutch. The easy-working, smooth, powerful hydraulic system has abundant capacity for all jobs including power lift, remote cylinder operation, and power steering. The 3-point hitch puts easy speed into "pick up and go" farming; exclusive Load-and-Depth Control makes possible top-notch tillage work in all conditions.

As you read the rest of this booklet, you'll see how well these tractors fit your specifications for power, economy, comfort, and ease of handling—whether you have a one-tractor, family-size farm or need a trustworthy, thrifty auxiliary tractor on large acreage.

Get the plus-value of a "compact"

It's easy to drive a "compact"—whether it's a car or tractor. And when you drive a "1010" you'll really enjoy the fast response to the throttle . . . the easy maneuverability . . . and best of all, the economy of a compact. It's a clean-lined, streamlined tractor—modern in every respect . . . a tractor you'll be proud to own. Give it a try. Call on your John Deere dealer soon, and ask him to arrange a demonstration at your convenience.

John Deere preceded the automotive industry in use of the word "compact," as the two pages of this 1010 brochure note. Even though not a large tractor, the 1010 was quite capable for its size, with 35 horsepower providing plenty of power for field and PTO work.

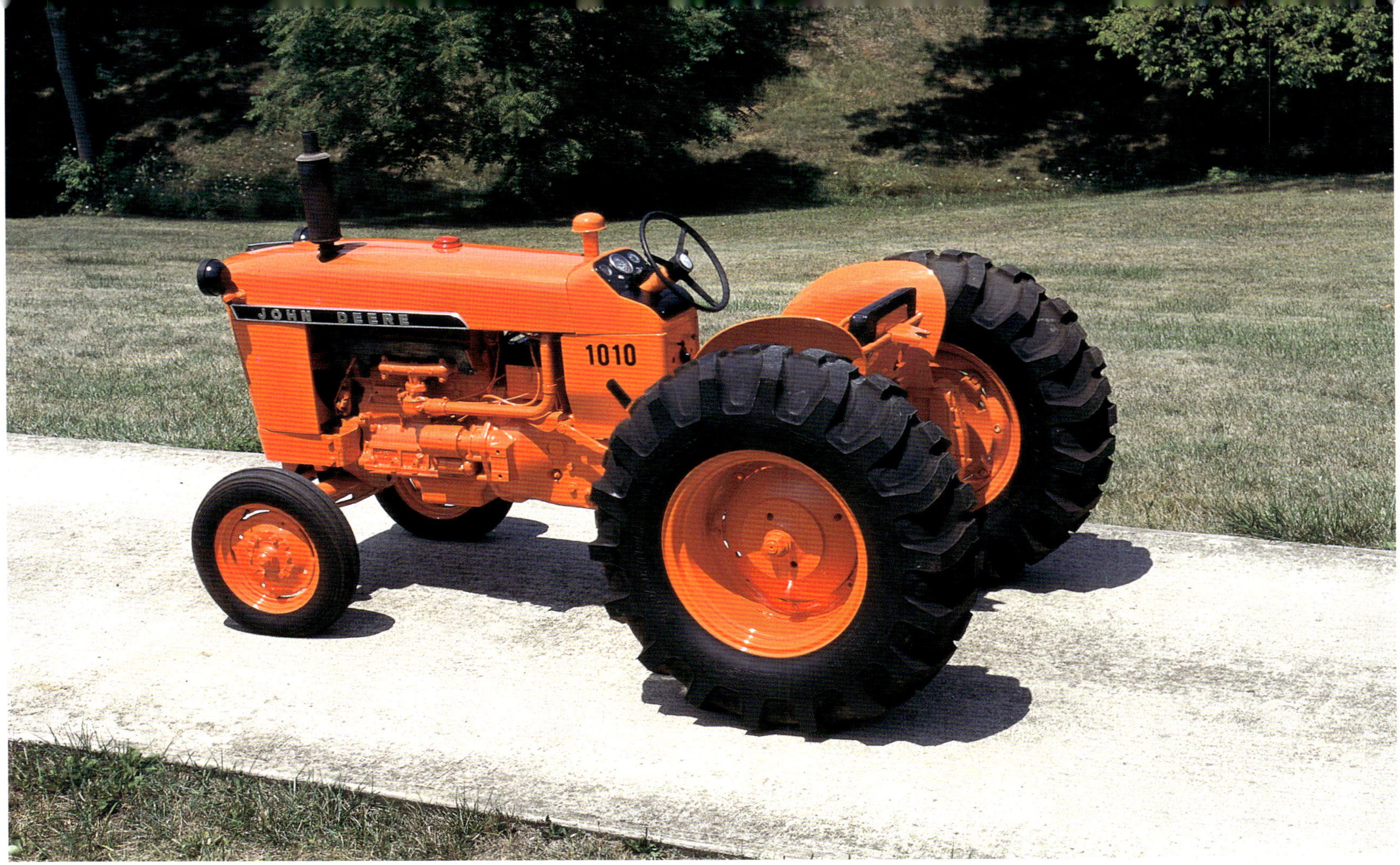

Because this 1963 1010 is an industrial model, it doesn't have power steering and a three-point hitch. Why the bright orange paint job? This color enhanced the safety of township employees operating it. The tractor cost $3,160 new in 1963. It weighed 3,923 pounds and had PTO horsepower rated at 36.1.

A New Generation of Earning Power

The John Deere 1010 SINGLE ROW-CROP Tractor

More usable power for all-round work . . .

tops for one-row cultivating

If you specialize in one-row planting and cultivating . . . yet need a tractor which can hustle a 2- or 3-bottom plow and similar equipment, and handle all types of tools for every farming job—the new "1010" Single Row-Crop is just right for you. It's great for tobacco, cotton, berries, vegetables, corn, flowers, and other crops. For precision work with a one-row cultivator, the front wheels are offset, and the seat adjusts easily to an offset position. Thus, you get a clear view of the row, straight ahead and down. You can see every plant, easily dodge those which are out of line. It has full 21-inch clearance, and a good range of wheel adjustments for all row widths.

The John Deere 1010 UTILITY Tractor

Down to earth in costs and design . . .

tops for dozens of jobs

This sure-footed "compact" offers you a whole new concept of utility power. With gasoline or Diesel engine, it delivers up to 35 horsepower for working such tools as 2- or 3-bottom plows with outstanding efficiency. And it can step along at fast pace with just about every type of equipment—drawn, PTO, 3-point and mounted. Low design gives it excellent stability for hillside use. The "1010" Utility is tops for specialized work in orchards, groves, and vineyards; on dairy, livestock, and poultry farms that do not require the power of a heavy tillage tractor; or as a second tractor on general farms. It's ideal for haying, hauling, and similar jobs.

Stressing "earning power," one of several New Generation tractor brochures displays large photographs of the two main 1010s, the single row-crop, and the utility tractors.

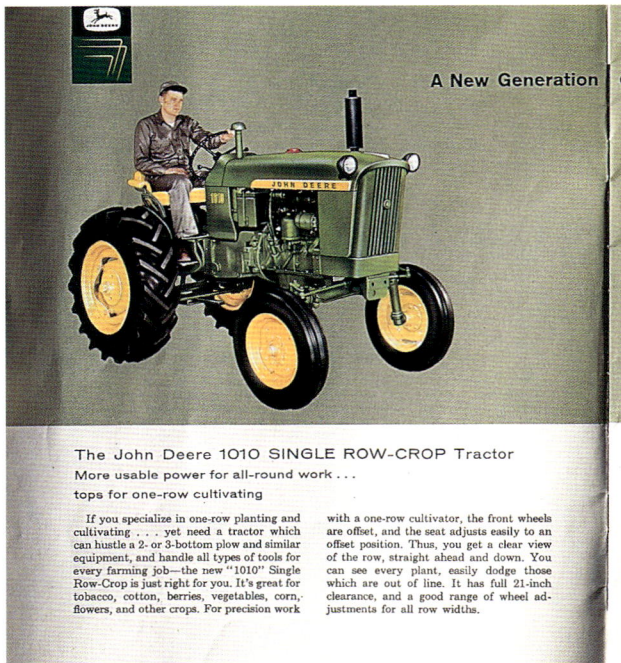

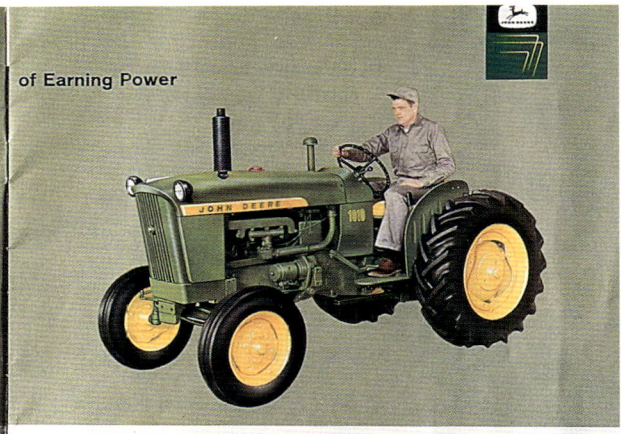

A bright yellow paint job indicated that this 1963 industrial 1010 was to be used only with a mower. It has no three-point hitch; however, mounted on the front of the crankshaft is a hydraulic pump that drives the mower and raises and lowers it.

The 1010 was designed to perform all the necessary farmstead jobs, such as scooping up manure from the lots with a front-end loader, as well as light fieldwork.

In addition to being the only completely restored 1010 crawler known to exist, this 1964 tractor is also unique in being gas-powered with a three-point hitch. Only 507—approximately 3 percent—of the 16,310 1010 crawlers built wore the Agricultural Green paint. Even fewer were gas-powered and mounted three-point hitches. The most common configuration was a diesel engine with front-mounted blade or bucket and a rear drawbar-type hitch.

INDUSTRIAL YELLOW LINE

The roots of Deere & Company industrial equipment extend back to the 1920s when the Waterloo Boy tractor was used to pull road graders maintaining Iowa country roads. In 1935, the Model DI, I for industrial, was introduced even though the Model D had found its way into industrial application 10 years earlier.

The yellow-painted line of I models continued virtually uninterrupted with the AI, BI, LI, MI, 830I, 840I, and others. The exception to the yellow color occasionally occurred when tractors left the factory painted orange. For example, MIs were painted "Nebraska Highway Orange" when sold to the Nebraska highway department for road maintenance.

Crawlers, painted green for agricultural and yellow for industrial usage, entered the line with Model D conversions in the early 1930s. In 1946, Deere & Company purchased the Lindeman Manufacturing Company of Yakima, Washington.

Lindeman had developed crawler conversions for the orchard versions of the GP and B tractors. Production of the crawlers and components was moved to the Dubuque plant after the Yakima facility was closed in 1954.

Today the industrial division of Deere & Company is a major manufacturer of yellow industrial equipment found at work in every corner of the globe.

The engine's centerline was "sacred ground," and for good reason. Once these dimensions were fixed and the tooling built, it would cost in the neighborhood of $70 million to replace the tooling if these dimensions should prove wrong.

These sacred dimensions resulted in the four-cylinder 3010 engine that developed 55-PTO horsepower in gasoline, diesel, or LP gas. The 4010 made its debut with a six-cylinder engine rated at 80-PTO horsepower in either gasoline, diesel, or LP gas.

Four alternative front-wheel options were available on the 3010 and 4010 tractors. Both tractors could be equipped as a row-crop utility or an orchard tractor. A hi-crop version was offered in the 4010 line. Also, both tractors came in Wheatland or standard models.

The 3010 and 4010 were replaced by the 3020 and 4020 in 1963. They remained in production until 1972, the last year of the New Generation tractors. Besides upping the PTO horsepower to 65 on

The 1964 1010 Crawler is reportedly quite easy to operate. Note the combination tachometer-speedometer. The gearshift is located on the transmission. The driver on a 1010 Crawler sat a half-foot higher than on the 430 Crawler that preceded it and could pull a four-bottom plow under most conditions.

A 1965 2010 Row-Crop with just 695 hours on its tach features the narrow front Roll-O-Matic. PTO horsepower is 45. Everything from tires to paint to seat is in original condition.

the 3020 and 96 on the 4020, the eight forward and four reverse speed Power-Shift transmission was introduced on these models. The 4010 and 4020 were extremely popular and ranked right up there with the Model D as classic John Deere tractors.

The 3010 and 4010 tractors and engines were manufactured at Waterloo, Iowa, while the 1010 and 2010 tractors and engines were a product of the Deere & Company plant at Dubuque, Iowa. Both of the latter models used Deere's sleeve-and-deck four-cylinder. The 1010 took gasoline or diesel and gave 35-PTO horsepower. The 2010, which took gasoline, diesel, or LP gas, provided 45-PTO horsepower.

Initially the 1010 and 2010 were built using existing tooling. When the 1020 and 2020 were introduced in 1965, the Dubuque plant installed new tooling for these models, which now sported most of the New Generation features of their big brothers.

The 5010 was Deere & Company's first tractor to break the "100 threshold" in both drawbar and PTO horsepower (this took place in 1963). Powered by a six-cylinder Deere & Company–built diesel engine, it was factory tested at 117-PTO horsepower and 100-drawbar horsepower. This unit was targeted at the wheat-producing Great Plains whose farmers needed to work vast acreages in a hurry. Standard equipment was a Syncro-Range transmission, deluxe seat, power brakes, power steering, wide rear fenders, and a dust shield. Cabs, factory-installed although not factory-built, were an option.

Two years later, the 5010 was upgraded to the 5020 model. With even more horsepower, it was a Wheatland mainstay until 1972. A Row-Crop version was available in 1967, the most powerful Row-Crop tractor available on the market at the time. All the options of its predecessor were also offered on this unit.

The 2510 joined the line in 1966 as the smallest New Generation tricycle Row-Crop tractor. Built at the Waterloo, Iowa, facility with either a Power-Shift or Syncro-Range transmission, it could be coupled to either a four-cylinder gasoline or diesel engine. The 2510 cranked out 55-PTO horsepower.

The 820 had the distinction of being built at both domestic and foreign Deere factories: Waterloo, Iowa, and Mannheim, Germany. It can also claim another distinction as one of two John Deere tractors to carry the model designation. The two-cylinder 820 diesel was produced at Waterloo, Iowa, from

2010 ROW-CROP TRACTOR
DUBUQUE

Here's New Generation Power that can broaden profit horizons of row-crop farmers—whether their acreage is small, medium or large. The new John Deere "2010" Row-Crop Tractor delivers up to 45 horsepower (PTO maximum factory-observed horsepower at maximum engine speed—2500 rpm.) With a John Deere 4-cylinder gasoline, Diesel, or LP-Gas engine, the "2010" hustles along with a 3-bottom plow, 4-row cultivator, hay tools of all kinds, mounted corn picker or cotton picker . . . in fact *all* kinds of 3-point, mounted, PTO, drawn, and integral tools. The "2010" Row-Crop offers four front-wheel assemblies and wheel treads for any crop or job. It's a low-budget tractor, too—top choice for the one-tractor farm or as a second tractor for big-scale operations.

Operator comfort and convenience have been stressed in designing the sleek, all-new "2010" Row-Crop Tractor.

QUALITY FEATURES

New, Variable-Speed Engines
The new John Deere-built variable-speed, gasoline, Diesel, and LP-Gas engines offer an exceptionally broad range of efficient engine working speeds . . . more flexible power matched to the job at hand.

New Syncro-Range Transmission
The smooth-shifting, flexible, new Syncro-Range Transmission delivers the most efficient speed for every job . . . speeds range from "creeper gear" to nearly 20 mph.

New Hydraulic System
A powerful hydraulic system offers a choice of single or dual rear rockshaft. A front rockshaft and a remote cylinder are also available.

Universal 3-Point Hitch
The "2010" Row-Crop 3-point hitch takes all types of equipment; provides a flip-of-a-lever choice of Load Control, Depth Control or exclusive Load-and-Depth Control. A new Quik-Coupler attachment lets the operator attach and detach tools from the seat.

"Live" Independent PTO
All types of power-driven machines can be driven by the "live" 540 and 1000 rpm PTO.

Front-End Assemblies
Interchangeable front-end assemblies include conventional and Roll-O-Matic dual front wheels, a single front wheel and an adjustable front axle to match crop and job requirements.

What made the 2010 Row-Crop tractor tick is fully described in this sheet. Note that the four front-end assembly options are considered so important that they're illustrated. This is another example of the detailed information that was provided by the technical writing staff at Deere & Company.

John Deere New Generation Tractors can put new earning power into any farm

1010
2010
3010
4010

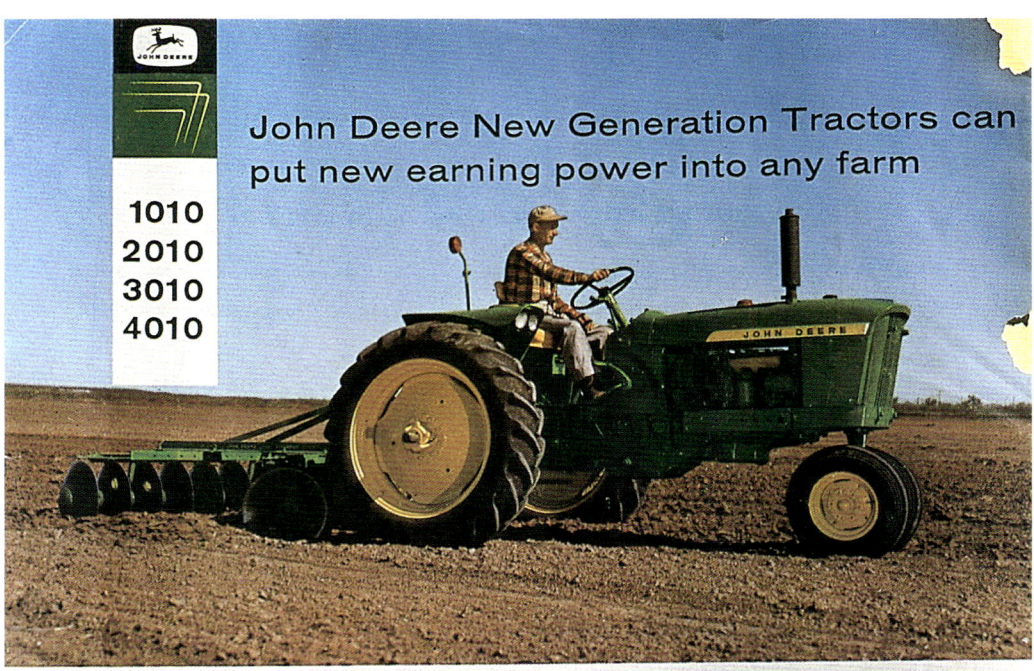

The "2010" Row-Crop Tractor with 414 Offset Disk Harrow.

15-

The New Generation of Power is a full line of tractors, all with the new variable-speed 4- or 6-cylinder engines. Take your choice of 35 h.p. "1010" . . . 45 h.p. "2010" . . . 55 h.p. "3010" or 80 h.p. "4010" with gasoline or Diesel engine, LP-Gas in the three larger sizes.

There are three Row-Crop models . . . two Row-Crop Utilities . . . two Standards . . . two Hi-Crops . . . Utility . . . Single Row-Crop . . . and Crawler.

Whatever type or size farm you operate, a new John Deere Tractor can give you New Generation Earning Power.

Ask your dealer to demonstrate the model that will put you years ahead in modern earning power

60-8-Dubuque

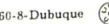

This page from a promotional brochure first appeared in August of 1960, at the same time the New Generation tractors were introduced in Dallas, Texas

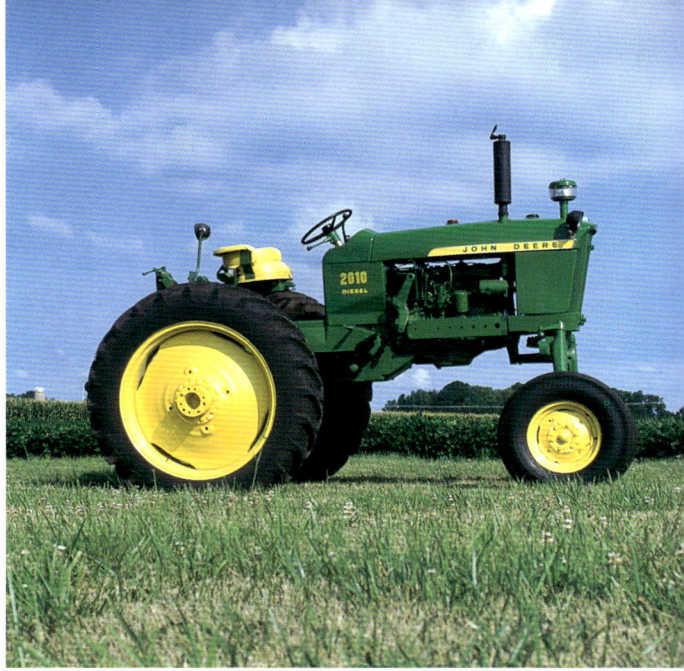

Before being restored with 400 hours of labor, this 1961 2010 Hi-Crop was basically broken down into parts stored in barrels. This tractor carries almost all of the important options—power steering, universal three-point hitch, and front rockshaft. There are 34 inches of clearance under the axles and 48 inches under the final drive housing for late season cultivation.

1956 to 1958, and 10 years later the New Generation 820 was powered by a three-cylinder 31 PTO horsepower diesel engine.

The following year (1969), the 2520 replaced the 2510 with a refined engine. This boosted the horsepower to 61 at the PTO. This unit stayed in the line until 1973.

Pulling smaller implements faster to accomplish the same amount of work was the underlying principle behind the 4000 that Deere introduced in 1969. Powered by the same engine as the 4020, it weighed 885 pounds less, and so pushed the previous limits of a high horsepower-to-weight ratio tractor. It featured the Syncro-Range transmission as standard equipment; however, the Power-Shift could be ordered as an option. Another option was the Roll-Gard and Roll-Gard canopy.

The advancements Deere made with the New Generation line followed from careful preparation. To make the transition from the two-cylinder line to the New Generation tractors, manufacturing was

It's logical that this 1966 2510 be considered a "10" series tractor. In this case, though, logic isn't correct. The 2510 was actually introduced after the first "20" tractors came out. The 2510 was the smallest New Generation Row-Crop. The 2510 was available with any of five different front ends, a gas or diesel engine, and either Synchro-Range or Power-Shift.

Deere & Company made just one 2510 Hi-Crop, gasoline, Power-Shift tractor—and this 1966 model is it. This is the kind of pot-of-gold tractor that is the dream of every collector searching for scarce models.

KEEPING THE SECRET

Little tests a person's character as being asked to keep a secret, even for a little while. Deere & Company asked dozens of hand-picked employees to "keep a company secret"—and to keep it for seven years. They did a remarkable job. In fact, this feat was later lauded as "the best-kept secret in industrial history."

In 1953, Deere & Company made the decision to replace its faithful two-cylinder tractors with an across-the-board line of new tractors. That same year, Deere & Company president and CEO, Charles Wiman, authorized an extensive research and development program to bring out this new line of tractors—tractors that had not yet been designed.

Nobody knew what these tractors would look like or what their specs would be. All that was known was that these tractors would be totally new and derived from a "clean sheet of paper."

It was a monumental decision and certainly a defining moment in Deere & Company history. Wiman put the entire company's financial future on the line at a time when its two-cylinder tractor line led the company in profits.

Management personnel, marketing people, product engineers, and production experts faced the job of peering years into the future. They were charged with determining what the marketplace would later demand in a rapidly changing agribusiness.

The same year Wiman made this decision to abandon the two-cylinder, he was diagnosed with a terminal illness and died in 1955. His son-in-law, Bill Hewitt, was chosen to fill the role as president and CEO of Deere & Company. It was during his watch that the New Generation of tractors moved from inception, to creation, to completion, to wide acceptance.

While nobody initially knew what would eventually take shape on the "blank sheet of paper," several things were apparent. To protect the sales of the company's current line of tractors, it was absolutely necessary that the impending change to multicylinder tractors be kept top secret.

It also became quickly evident that personnel couldn't carry on the new development work simultaneously with their current duties. Also, if secrecy was to be maintained, the project needed to be moved to a remote location where only those people involved with the project would have access to the ongoing developments. Every detail concerning the New Generation was released strictly on a need-to-know basis.

An empty supermarket was leased in downtown Waterloo, Iowa, in 1953. It became the engineering center for the initial work on the tractor design. Veterans of the project still refer to the facility as the "meat market." A year later, the first building was constructed on the site that became Deere & Company's new Product Engineering Center (PEC).

The rest of the facility wasn't occupied until 1956 when the New Generation project was moved to the PEC. Here work continued. The design for Deere & Company's new tractor was pretty well set by 1958 and in production before the end of 1960.

Seven successful years under wraps, the competition might have suspected something was afoot at Deere & Company; but that's all they had—suspicions.

Keeping the secret required some rather innovative measures, such as making sure the tractors were always covered or enclosed while being transported to test facilities. Test models were painted a color other than John Deere Green and John Deere Yellow. Side panels were bolted onto the tractor to camouflage their distinctive profiles.

The two-cylinder engine had a distinct exhaust note that earned it the nickname "Johnny Popper." The exhaust sound of a four- or six-cylinder engine destined for a New Generation was distinctly different.

When engines are tested in the test cells at the Deere & Company PEC, the exhaust is vented through the roof of the building. Engines are tested right up to the red line in these test cells and run for hours. The problem: A two-cylinder working under load can be heard clearly at least a mile away.

Apparently no industrial spy or competitor ever picked up on this obvious clue and investigated what was producing the different exhaust sound hour after hour. It may be only a "war story," but there's a legend at Deere & Company that a recording was made of two-cylinder exhaust noise—then it was played full blast through a loudspeaker when the new four- and six-cylinder units were tested in the PEC cells.

Bottom line: Deere & Company kept its secret for seven years, and then surprised the industry with a line of tractors that redefined the standards for years to come.

---

stopped at the Waterloo plant for approximately six months while the new tooling and assembly lines were set up. A large inventory of two-cylinder models was built and stored to provide dealers with tractors until the New Generation series came on line.

Deere also began to restructure its tractor program within two years after the New Generation line was launched, so that all manufacturing facilities were "reading from the same page."

As Harold Brock, retired director of the PEC, explained the thinking: "Our plan was to design a 'worldwide' tractor at Waterloo. Dubuque made one kind of tractor, while Mannheim, Germany, was making tractors of another design. And both of these differed from the concept of the 3010 and 4010. We needed a tractor that would satisfy the world.

"Our purpose was to determine what kind of a product would be best for markets on a worldwide basis. In Germany, if you machined a part from our U.S. blueprints, it would turn out to be the opposite hand of it, or reversed. So we realized we needed drawings that wouldn't have to be converted by any factory. We set out to make drawings that wouldn't have to be redrawn, and from which any Deere & Company factory in the world could make the part. This was the foundation of our worldwide concept of drawings."

The concept was ideal, but even Deere & Company sometimes had to be satisfied with a little less. Mike Mack, retired director of the PEC who succeeded Brock, commented: "It's never going to happen 100 percent, but the object is to come as close as you can. Sure there are certain standardizations that can be affected just from a functional point of view. But, there are a lot of differences that have to remain. The people in Europe have different climatic conditions and cultures, for example.

"The row spacing is different in foreign countries, so there are different requirements in tread width. Foreign governments have different emissions and noise-level standards. All this brings about requirements that are unique to that particular trade area.

"All these varied requirements must be considered in the planning stages by the staff at the PEC. But, at the same time, the staff is striving for the most commonness for every market."

Deere & Company engineers were also working on two other important programs during the 1960s. One was ROPS. The second was the

Looking almost showroom new, this is an impeccable 1969 2520. This was the year the series was introduced. A new designation of 2520 resulted when the 2510 received an additional 6 horsepower.

"Unique" barely describes this precisely restored 3010 Row-Crop diesel. It was the first one of all the 3010 series tractors from Deere's Waterloo, Iowa, tractor plant in 1960. Various front-wheel options included adjustable front axle, conventional dual front wheels, regular or heavy-duty single-front wheel, or regular or heavy-duty Roll-O-Matic "knee-action" duals.

Only in this particular model, the 3010 LP, was the LP gas tank situated where the gasoline tank normally was positioned in the other models. While annual sales of the gasoline and diesel models were usually in the thousands—with one 3010 diesel model topping out at 7,171—the most of any LP 3010 model made in a year was just 976.

Although the 3010 was a popular success, the 3020 replaced it just three years after it premiered on U.S. farms. A more powerful tractor, the 3020 offered the Power-Shift transmission and provided more power. Now the operator could shift with a single lever into any of eight forward speeds or four reverse speeds without clutching, and even while moving. Although this particular 1966 3020 originally mounted a gas engine and Power-Shift transmission, the engine used too much gas and was relatively low in horsepower for its weight. Its owner built an adapter plate, then installed a John Deere industrial tractor diesel "Dubuque" engine. This increased horsepower from 70 to 105, plus it also enhanced fuel efficiency. As a result, the modified 3020 applied almost the same horsepower to its three-point hitch as a late 4020 tractor.

When new, this 1965 3020 Wheatland was the mainstay of a Canadian farmer in a prairie province. How it differed from its fellow 3020s: narrower front end, larger fenders, drawbar instead of a three-point hitch, and the driver got on from the rear rather than from the side.

Coffee shop debates still center on which was the better tractor, the 3010 or the larger 4010, which combined 80-PTO horsepower strength with practical symmetry and an unexcelled power-to-weight ratio. This particular tractor was the first 4010 to move down the assembly line in 1960.

Sound-Gard cab. The first was designed to protect the operator from injury due to a vehicle rollover, while the second not only protected the operator from a rollover, but also from heat, cold, noise, and dust. When the Sound-Gard cab was finally introduced in 1972, it offered operators a comfort level equal to that of most automobiles of the era.

These innovations, and others encouraging operator comfort, were what former CEO Bill Hewitt called the "human factor." They were matched with improved styling to make John Deere tractors even more attractive to prospective buyers.

Hewitt explained the importance of styling. "Let's assume that two competitive tractor companies each have the horsepower and capacity that you as a farmer want," he said. "Say both tractors are theoretically equal

Manufactured in 1962, the 4010 Diesel Row-Crop was built only with Synchro-Range transmission. What distinguished the row-crop model was that it could have either a narrow front end or a wide front end. Both the rear and front wheels were adjustable for width. The narrow front-end model was favored by farmers who mounted a corn picker on it, because the tractor could travel between the rows without knocking down any corn.

Every important detail about the 4010 Row-Crop tractor is itemized on this sheet about the advantages it provides, including being able to pull an eight-row planter or five-bottom plow.

29

What goes around, comes around: The shop mechanic at the Iowa John Deere dealership who prepped this tractor for delivery, the first 4010 Diesel to be sold by the dealer in 1960, now owns both the dealership and the tractor, which he found and restored.

The 4010 Diesel Wheatland is pretty much similar to a 4010 row-crop with these exceptions: fixed-width, wide front end; smaller, 18.4x30 tires; and wider, rounded fenders. The rounded fenders gave the perception that they kept more dust and dirt off the driver. Destined for the Great Plains market and built strictly for pulling tillage equipment and grain drills, there was no three-point hitch or PTO.

in mechanical performance, serviceability; in short, they're similar in every facet. But one of these tractors is a lot better looking than the other. They're both in the same price range. So which tractor are you going to buy—especially if your wife's along with you?"

The task of making the New Generation tractors "a lot better looking" fell to industrial designer Henry Dreyfuss and his staff. This wasn't Dreyfuss' first effort for Deere & Company because he'd had input on Deere's tractor design since the 1930s; however, it was the first time he'd been in on a design project from the ground up, something he relished.

One feature that received special consideration was the seat. The Dreyfuss firm enlisted the services of Dr. Janet Travell to design an ergonomically correct seat for the new line of tractors. "She knew just where the armrests should be, the dimensions of the armrests, and the amount of softness or firmness in the armrests," Hewitt noted. "She also had the engineers design a spring-loaded seat, so that the seat was not rigidly attached to the hard part of the tractor."

For several years following the New Generation's introduction, Doctor Travell's name appeared

Developing 96-PTO horsepower, the 4020 was the mainstay of the "second generation" of the New Generation tractors. It had almost everything a farmer could want: adequate power, ease of operation, reliability, and the famed John Deere service and parts availability.

What farmers really liked about the later 4020s was that they were standard equipped with Power-Shift transmissions. This enabled them to be shifted on the go and under load, and without any clutching.

Rather than being flat, this 4020 steering wheel has a definite dish to its center, which brought the rim nearer the driver. This identifies it as a 1971–1972 model. The right-side hydraulic lever console was standard in 1969–1972.

31

frequently on the front pages of the nation's newspapers. As a back specialist, she became the personal physician of President John F. Kennedy, who endured back problems. It was her suggestion to install the famous rocking chair in the Oval Office.

The last of Deere's New Generation models would continue into the early 1970s, further refining 1960s technology. Three new models were added in 1971: the 4320, 4620, and 7020. The 4320 was sometimes called a "hopped-up" 4020 or the "super 4020," thanks to a turbocharger boosting the six-cylinder diesel engine's power to 115-PTO horsepower. This tractor featured most of the same standard equipment and options as the 4020, including power front-wheel drive.

The 4620 added another engine feature to Deere's six-cylinder diesel: intercooling. When the art of turbocharging reached a plateau, engineers added a component that cooled the air before it entered the combustion chamber. With this feature the 4620 was capable of 135-PTO horsepower.

continued on page 36

This 4020 Diesel carries serial number 65001. This particular tractor was the first Power-Shift tractor ever manufactured by John Deere when the 4020 replaced the 4010 in 1964. Its tachometer has recorded only 3,000 hours since then.

This unusual 1966 4020 LP, narrow front tractor that carries a large front-mounted LP tank is the only homely New Generation tractor built. A three-point hitch and large rear tires are features of this particular unit.

32

A 1965 4020 Diesel such as this one could be bought with optional dual hydraulics, Power-Shift, and wide front end. The 4020 replaced the 4010 in 1964 and was one of the most popular John Deere tractors ever built. It remained in the line through 1972 and was available in gasoline, diesel, and LP gas coupled to either a Syncro-Range or Power-Shift transmission. Whatever configuration the customer opted for he could count on horsepower in the 91 to 96 range. The 4020 was offered as a Standard, Row-Crop, Row-Crop Utility, or Hi-Crop model. A farmer was able to "customize" his new tractor to dovetail it with his needs. For example, he could purchase a new 1965 4020 Diesel, such as this one, with optional dual hydraulics, Power-Shift, and wide front end.

Here's Dependable John Deere Service

Servicemen
who know their Business

You can place full confidence in a John Deere Dealer's Serviceman. He's skilled . . . experienced . . . knows equipment inside-out.

Trained in servicing methods recommended by John Deere, these mechanics know what to do and how to do it; they keep up to date on methods, changes, and new machines.

This assures quality work . . . quickly and efficiently done . . . *the first time* . . . and at the lowest cost. Remember . . . you'll get your money's worth with John Deere Service!

Parts Departments
well stocked to meet your needs

Having parts when you need them—that's the true test of good parts service!

That's also the aim of your John Deere Dealer. His parts department keeps a good supply of JD Parts at all times. Should he be temporarily out of stock, additional parts can be secured quickly . . . often overnight.

Most important, you'll get only *genuine* John Deere Parts. They're made right . . . fit properly . . . last longer . . . for continued, dependable John Deere performance.

Your John Deere Dealer - Home of Quality Service, Genuine John Deere Parts

hO

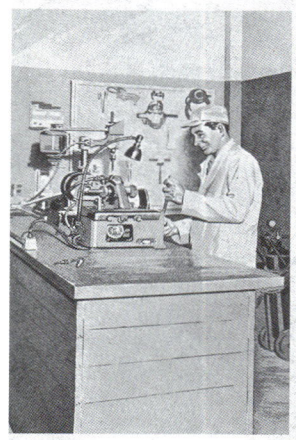

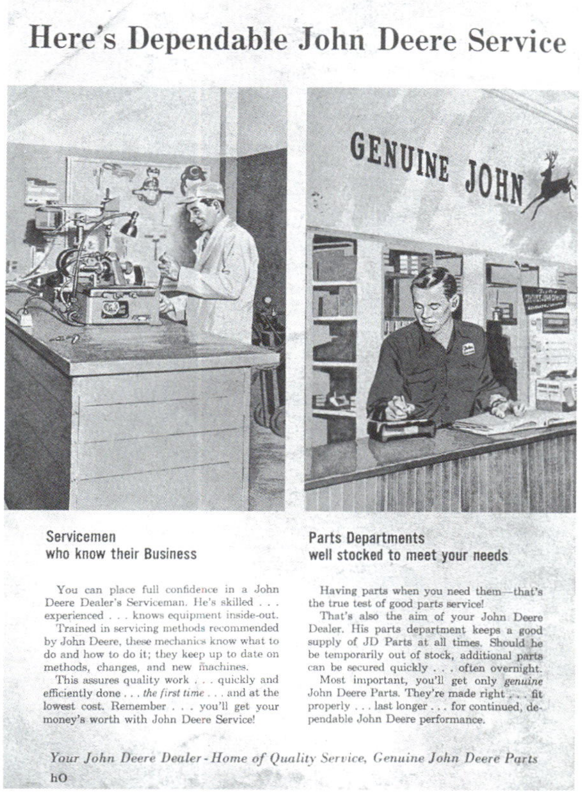

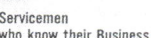

Many farmers believe that "No parts and no service, no tractor!" The real strength of Deere & Company lies in its unsurpassed service and parts availability. Although this important foundation had been set with earlier generations of tractors and equipment, the concept received even more emphasis during the 1960s.

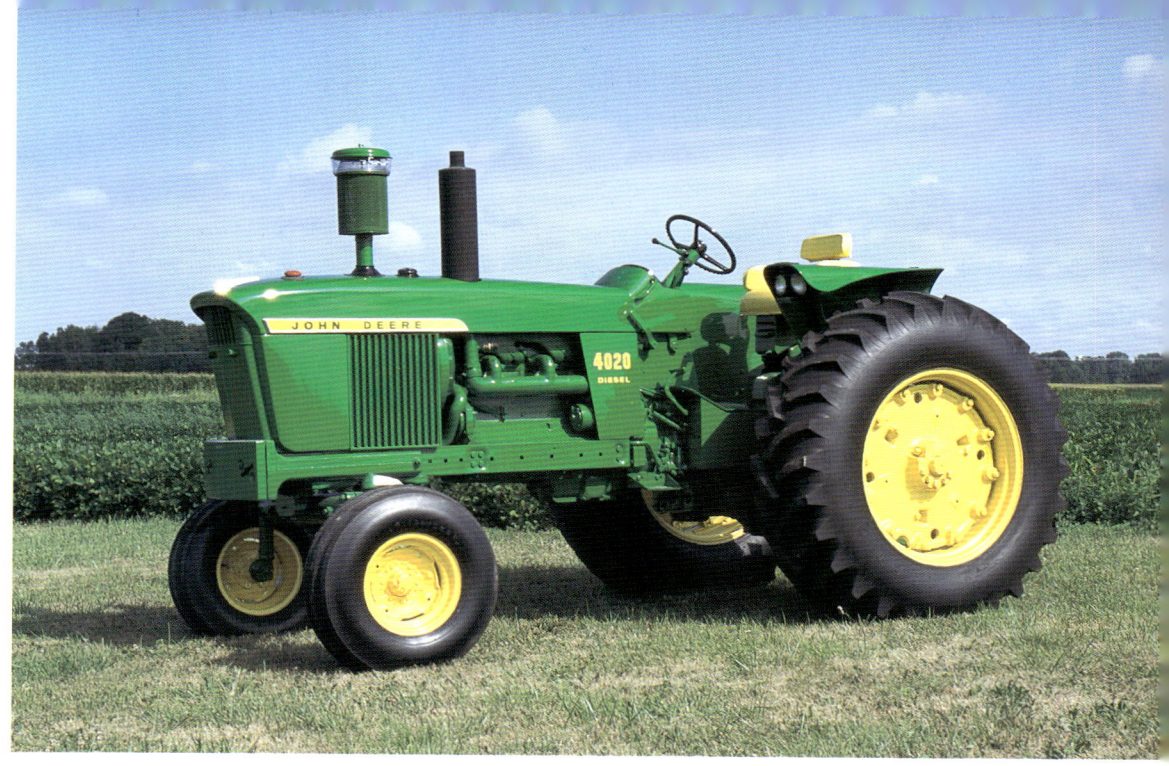

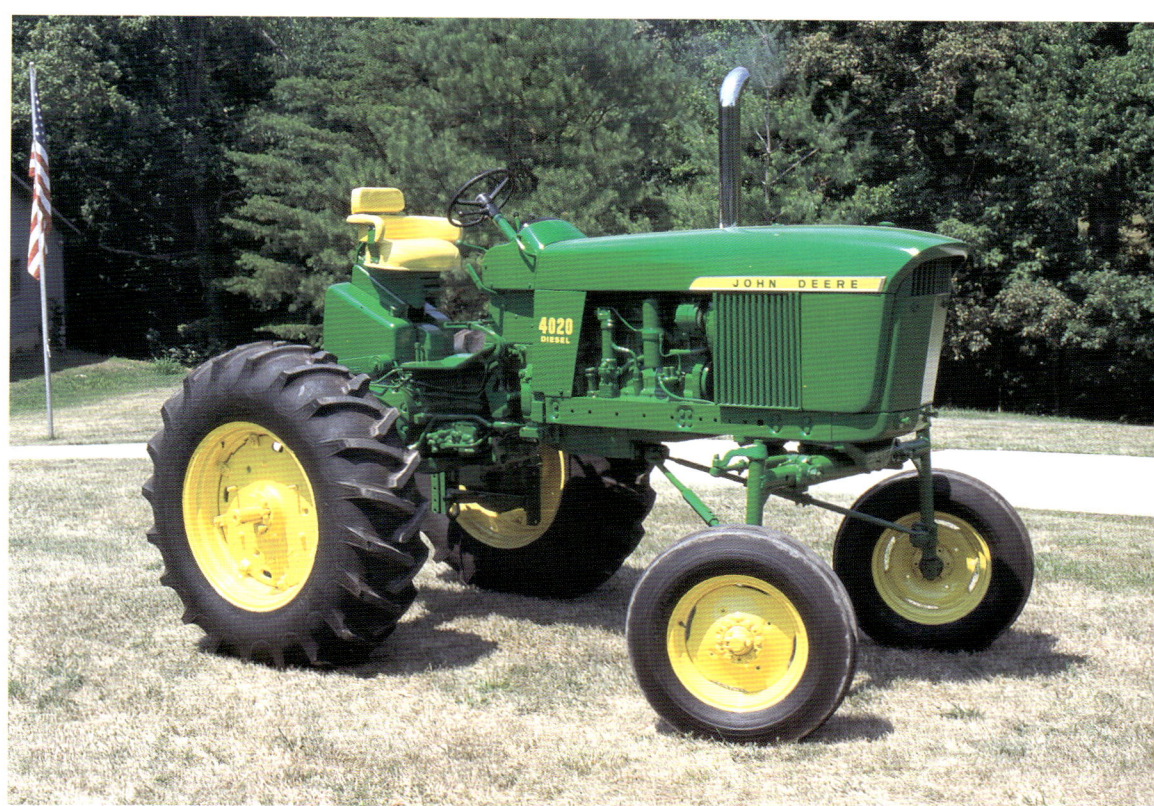

Powered by a 96-horsepower engine, this 1965 4020 Diesel Hi-Crop was designed and built for cultivating sugarcane in the South and truck crops in California. When new it sold for $11,370.

Extremely rare, and seldom seen in the Midwest, is the power-front wheel assist of this 1969 4020 Diesel. Most such tractors were shipped to the Western states, especially Idaho, where they were used to plant, cultivate, and harvest potatoes. Now this one is used for transporting big round bales.

Like most of its brethren, this 4020 Diesel powering a grain auger is still working hard and lookin' good. The 4020 has been—and continues to be—a mainstay on thousands of U.S. farms. The 4020 was powered by a Deere-built six-cylinder, variable-speed, wet-sleeve, valve-in-head engine designed to operate between 800 and 2,200 rpm.

Although most late 4000s and 4020s were diesel-powered, this 1969 tractor carries a gasoline engine. The 4000 and 4010 mounted the same six-cylinder engine, while the 4000 shared the same rear end with the smaller 3020. The 4000 also used a less costly transmission and lacked some of the finishing touches of its stablemates. It was targeted at the farmer who wanted more power, but in a smaller tractor that wouldn't be overloaded. The 4000 was an attempt by Deere & Company to fill a gap in its lineup of tractors in order to better compete against the International 706 tractor.

Few of the 4000 Diesels made were exactly like this one: The 4000 was a tractor designed for the economy-minded, yet this 1969 model uses the more expensive Power-Shift transmission option. It has only 2,347 hours on its tachometer, and its paint is original. The tractor came with absolutely no frills, such as battery covers or sheet metal around the seat to cover the hydraulics.

A gasoline model, this 1969 4000 94-horsepower, gasoline-powered tractor has two fewer horsepower than the comparable diesel 4000, and less torque. Deere & Company produced just 256 4000 gas tractors with both Synchro Range transmissions and Roll-Gard canopies.

This 1969 4000 Low Profile Diesel uses a fixed-tread front axle from a 3020 Row-Crop utility tractor factory-installed in the short-wheelbase position. Introduced in 1969, the 4000 used the same engine as the 4020 but weighed 885 pounds less giving the 4000 a high horsepower-to-weight ratio. The concept was to pull smaller implements faster to accomplish the same amount of work. The customer could have either the standard Syncro-Range transmission or the optional Power-Shift.

35

D-DAY IN DALLAS

No expense was spared by Deere & Company to make the introduction of the "New Generation of Power" impressive and effective. The extravaganza was two years in the planning and cost in the neighborhood of $1.25 million in 1960 dollars.

The event combined the brilliantly staged choreography of a Broadway show with the logistics of a military invasion. It was a departure from Deere & Company's traditional method of introducing a new product line. In the past, new products and key personnel were taken to regional sites around the country for introductions and demonstrations.

This time, Deere & Company felt it was only proper to showcase this "new-from-the-ground-up" line of tractors and equipment on a much grander scale. So, the decision was made to bring Deere & Company's entire worldwide staff of distributors, dealers, and guests to a single location for one big day of introduction and celebration.

The place was Dallas, Texas, and the date was August 30, 1960. Deere & Company provided air travel for most invited guests. The air armada began touching down at Love Field, Dallas, in the early hours of August 29, 1960.

By the time the last flight was on the ground, 5,000 Deere & Company passengers had participated in what was called the greatest airlift of its type ever attempted. These distributors, dealers, and guests were joined by an additional 1,500 people who had arrived by bus, railroad, or automobile.

Bus transportation ran like clockwork shuttling guests to motel and hotel rooms near the Memorial Auditorium and the Livestock Coliseum on the Texas State Fairgrounds. Deere & Company President and CEO Bill Hewitt recalled with satisfaction that there was only a single incident of lost luggage.

Beginning at 10 a.m. the next morning, the New Generation line rolled out for everyone's inspection via film. People could see and have their appetites whetted, but they still couldn't touch. Yet. That would have to wait until after the Texas-sized lunch. The Walter Jetton catering service had 6,500 box lunches waiting for the hungry visitors at high noon.

Meanwhile, Hewitt and other VIPs departed for downtown Dallas and the prestigious Neiman-Marcus store. On the first floor of the store, right next to the jewelry display, a 3010 Diesel row-crop New Generation tractor was unveiled to the expectant spectators. The tractor was adorned with "approximately a million dollars worth" of diamond jewelry, according to Hewitt.

After this bit of showcasing, it was back to the Memorial Auditorium. The "fed and watered" crowd was then bused to the Livestock Coliseum where the impatient visitors could finally get a chance to see the tractors in real life.

Here the tractors were paraded through the arena with "cowgirls" carrying signs identifying each model. The review was greeted with continuous cheers and applause until the last tractor rolled past.

Finally the doors were opened and the anxious crowd moved out into the outdoor exhibit area to see, touch, and examine the tractors. The area contained 15 acres of product that included 136 new tractors and 223 implements—almost the entire Deere & Company agricultural and industrial lines.

The 5010 tractor described in this sheet is called the "most powerful standard tractor on the market." It provided a hefty 100 horsepower at the drawbar and 117 horsepower at the PTO. Coming into the Deere line in 1963, the 5010 was aimed at the wheat-producing Great Plains. It topped the Allis-Chalmers D-21 Diesel, also a 1963 offering, by 14 horsepower at the PTO and 6 horsepower at the drawbar. The Massey-Ferguson 97 Diesel (M-M G 705), another 1963 offering, weighed in at 101-PTO and 93-drawbar horsepower.

Continued from page 32

Deere & Company's second attempt at manufacturing four-wheel-drive tractors was the 7020 introduced in 1971—more than 10 years after the 8010 was showcased. In 1969, Deere reached an agreement with Wagner to sell their four-wheel-drive units carrying the John Deere badge and colors. In the interim years, Deere's tractor line didn't offer an articulated four-wheel-drive tractor.

Some owners considered the 7020's six-cylinder diesel engine a bit underpowered at 145-PTO horses. This was increased to 175-PTO turbocharged horsepower in the 7520 version, which came on-line in 1972.

How Deere reentered the four-wheel-drive market with the 7020 is explained by Harold Brock, retired director of the PEC: "What we did with the 7020 was to take the 4020 and bisect it, and then put an engine in the middle. By doing this we took advantage of high-volume production components, which brought the price down. When we put our

Back in 1963 when it was shiny new, this 5010 was the largest, and with 117 horsepower, the most powerful, row-crop tractor built anywhere in the world. Its power was so appreciated that some 5010s worked outside of agriculture and carried industrial yellow paint. The 5010 is given much of the credit for initiating the trend to ever larger tractors and equipment that continues today.

This 5010 is cutting a wide swath as it operates a chisel set well in halfway up its shanks and a following packer. The dual rear wheels limited soil compaction. Although fairly heavy, the tractor used a 133-horsepower six-cylinder engine of 531 cubic inches.

Talk about hybrids! Mounting a Cummins 855 Diesel putting out 400 horsepower and loads of torque, this 1966 5020 now carries a special 5050 model decal that the converter-restorer had made to commemorate the one-of-a-kind adaptation. It also features a cab from a 4440.

high production parts together we found that we could sell our tractor for approximately $13,000. That put us in the four-wheel-drive business."

The last year of the New Generation tractors, 1972, saw the introduction of the 6030. It carried the 30 Series designation and stayed in the line for five years. It was equipped with a 531-cubic inch, valve-in-head, turbocharged, intercooled six-cylinder diesel engine that was developed for the 7520 tractor. It rated 175-PTO horsepower. And, for the first time, Deere & Company offered a choice of engines, either the naturally aspirated version or the turbocharged and intercooled model.

THE TRADITION CONTINUES

"Generation II" John Deere models were introduced in 1972. They continued to improve the innovative features of their predecessors.

The restorers believe this 1960 8010/8020 is one of only two mounting a cab, a damaged 7020 cab that Deere & Company converted to fit. This particular unit was first painted yellow and used as a "yard bird" tractor at the Waterloo, Iowa, plant. As such, it was the last 8010/8020 to leave the factory. Powered by an inline six-cylinder Detroit Diesel engine, this tractor has never been overhauled. Three of its four tires are original.

Deere & Company shipped this unique, first-of-its-kind-in-several-ways, tractor to Montana covered with canvas to prevent other manufacturers from seeing what John Deere was doing. It was never sold to a farmer and later returned to the Waterloo, Iowa, plant where it was used for factory yard work. The 8010 is the first four-wheel-drive tractor manufactured and marketed by the company. This particular unit is the first ever to be put together, and it's the only 8010 not going back to the factory on a recall to replace the transmission and clutch to be returned as an 8020 model. It's reported that farmers were reluctant to buy the 8010/8020. They thought the monstrous tractors were too big, would burn too much fuel, and cause too much compaction. John Deere had to resort to renting many of the 8010/8020s before they could eventually be sold. Yet farmers found it intriguing to see an 8010/8020 roar through a field turning soil with a fully mounted eight-bottom plow.

A giant of a tractor, its steering wheel is more than 8 feet off the ground, and the tractor is close to 10 feet long and weighs almost 10 tons. Power was 215-engine horsepower and 150-drawbar horsepower. Only two 8010s were ever sold east of the Mississippi River.

The 1980s ushered in a farm economy that rivaled that of the 1930s: stock market woes, the depression, and dust bowl. Deere & Company survived intact—a singular accomplishment among the major agricultural machinery manufacturers.

As more than one farmer has reasoned, "Deere & Company is the only major farm equipment company that hasn't gone belly-up or had to merge. When I bought my new John Deere tractor I planned on using it for at least 10 years. A tractor nowadays is an expensive investment and can easily run into six figures.

"It was important to me to have some assurance that if I needed parts and service they would be there. So why gamble with another make? I figured I didn't have any worry buying a John Deere!"

The scope of the company has expanded to include, besides a full line of agricultural tractors and equipment, construction equipment, credit division, lawn, golf and turf care products, insurance, health care, and worldwide parts distribution.

Could John Deere have envisioned what his namesake company would become? Probably not. Would he be pleased and gratified? Yes. Would he be astounded? Without a doubt.

Today, the enterprise he started in a small shop in Vermont has grown into a massive global corporation that conducts business in 160 countries and employs 33,900 people. It's a company that in 1998 showed its thousands of stockholders a net profit of slightly more than one billion dollars.

39

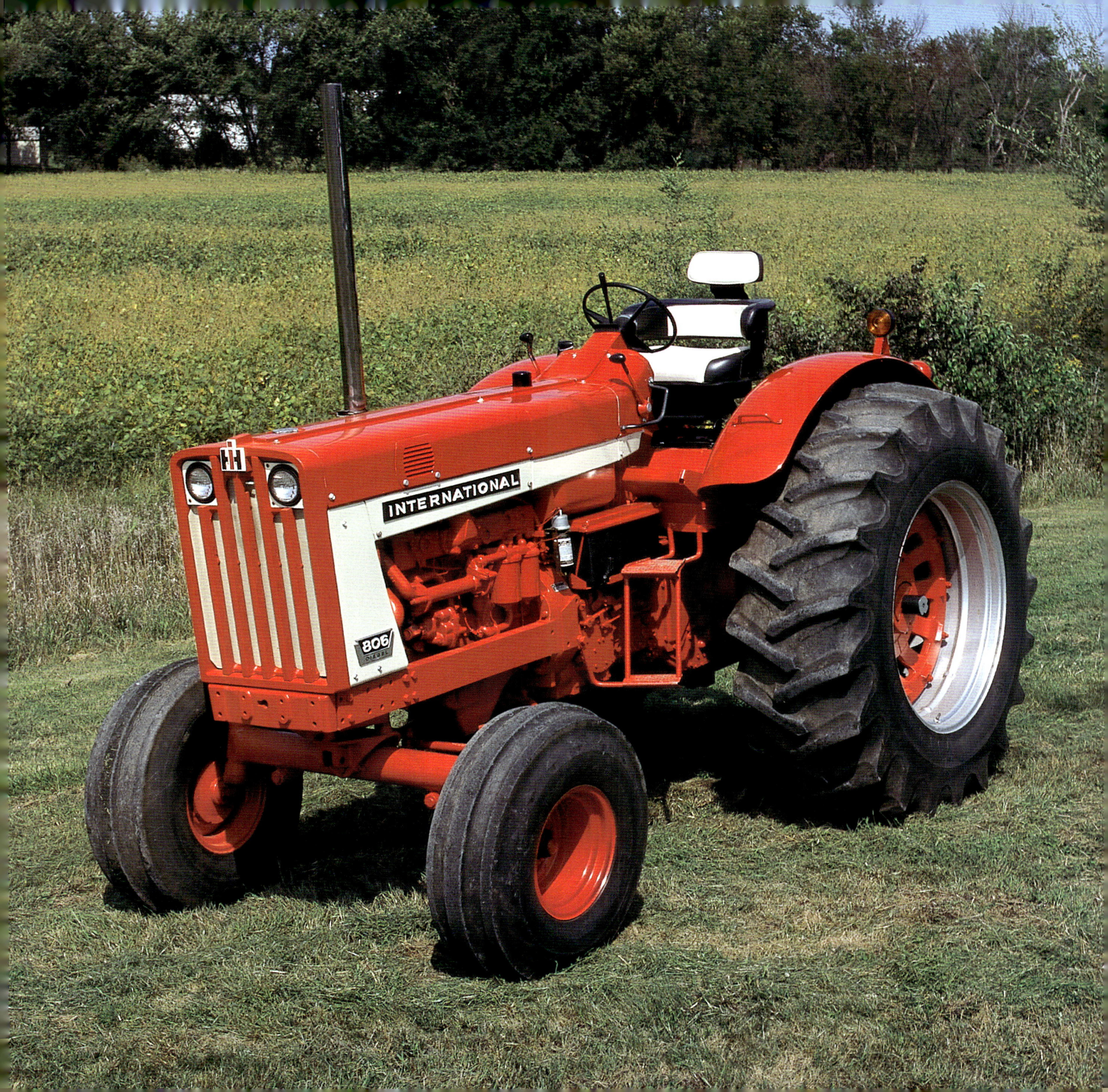
INTERNATIONAL
806

CHAPTER TWO

INTERNATIONAL HARVESTER COMPANY

Locate the headwaters of the giant International Harvester Company watershed and you'll find Cyrus Hall McCormick. The son of Robert McCormick Jr. and Mary Ann Hall, McCormick was born February 15, 1809, the first of eight children.

The McCormick family holdings at Walnut Grove, Virginia included some 500 acres of land with grist mill, distillery, hemp brake, sawmill, lime kiln, and assorted livestock. This made the family of "some means" on the expanding frontier.

With the help of Jo Anderson, a family slave, McCormick built his first reaper, which he demonstrated to the public in July 1831. Unbeknownst to McCormick, another man, Obed Hussey of Ohio, had developed a similar invention about the same time. When McCormick heard about it, he filed for a patent and told Hussey to stop producing the machine McCormick had invented. The rivalry led to a contest between the two reapers in 1843. McCormick was declared the winner.

In 1847, Cyrus moved to Chicago to build a factory and establish a business, securing valuable help from his brothers Leander and William. Following a series of partnerships, some more harmonious than others, the company began to flourish. By 1879, it became the McCormick Harvesting Machine Company. Cyrus died in 1884 and his son, also named Cyrus, took over with help from his mother, Nancy, better known as Nettie. Though the company and its profits continued to grow, competition would bring change.

In 1902, the International Harvester Company (IHC) of Chicago, Illinois, was formed by the merger of McCormick Harvesting Company, Deering Harvesting Company, Plano Harvesting Company, Milwaukee Harvester Company, and Warder, Bushnell, and Glessner Company (Champion harvesters). The merger brought 85 percent of the U.S. harvester business under one flag. The merger, however, did not involve large consolidation of operations. The retention of the McCormick, Deering, and Plano names with similar model lines led to duplicated effort and incompatible parts across the organization. Internal competition, as prevalent as cooperation, would lead to internal problems at International Harvester for decades to come.

Still, the massive company was a formidable market force. By 1909, IHC's assets of $172.7 million placed it fourth in size among all U.S. corporations. Only U.S. Steel, Standard Oil of New Jersey, and the American Tobacco Company, all industrial amalgamations, perched higher on the assets value tree. More than 40,000 dealers selling either McCormick, Deering, Plano, or the other brands represented the new company's marketing system around the country.

IHC commissioned its first tractor in 1905, releasing its primitive Type A a year later. The 10,500-pound machine featured a 20-horsepower, one-cylinder engine with an open crankcase and "total loss" oiling. The driveline was engaged by moving the engine back and forth on rollers.

International Harvester was one of the dominant makes of the 1960s, with innovative new machines and an almost unmatched brand loyalty. Despite this, the rise of John Deere would see IH move from its more customary position as the leading manufacturer into the number two slot. This 1966 806D has a 361-cubic inch engine that puts out 94 horsepower. The 806 was produced from 1963 through 1967.

41

Farmers were curious about the new technology, but wanted smaller, more versatile machines. In 1914, IHC responded with the four-cylinder Titan 12-25 and the single-cylinder Mogul 8-16. Mogul sales alone broke 10,000 within the first two years. By 1920, some 200 companies offered tractors to U.S. farmers, with total production exceeding 200,000 units. Though IHC sales were strong, Ford led the market, followed by IHC and Case.

To capture the hearts and minds of farmers, IHC engineers worked to create "the perfect tractor." They came close with the Farmall, introduced in 1924. In a big way it was responsible for the company's record profits of $37 million in 1929. The value of IHC common stock soared to a record $142 per share in that year.

In 1931, 100 years after McCormick demonstrated his first reaper, there were 31 IHC factories

The 135-cubic inch four-cylinder International Harvester–built gas engine in this 1960 340 Hi-Crop provided 35 horsepower. The series was manufactured from 1958 through 1963.

producing an ever-widening array of goods for a consumer market that was becoming increasingly cash poor. Drought and ensuing erosion had produced the "dirty thirties," compounded by the Great Depression. IHC's sales dropped more than two-thirds, but the company held on.

Despite hard times nationally, IHC introduced many new models in the 1930s, including the F-30, F-20, W-30, F-12, W-12, and F-14. In 1934, the W-40 was added to the line, with a six-cylinder diesel engine and 35 to 50 horsepower. The WD-40 was the first U.S.-built wheel tractor to use Rudolf Diesel's engine.

The company made styling changes as well, hiring Raymond Loewy, industrial designer of the Studebaker and much later, NASA's Skylab, to give IHC's tractors and crawlers a more modern and streamlined appearance.

Small, yet quite agile, this 1960 T-340 crawler obtained its 36 horsepower from a gasoline engine of 135 cubic inches. It was used for tillage work. Many such crawlers also carried dozer blades for light dirt moving work.

Made in 1961, this 660 Wheatland came equipped only with a hand clutch—no foot clutch—and a torque amplifier. It was available with a gas, LP, or diesel engine. The six-cylinder diesel engine in this particular tractor was rated at 82 horsepower and could power the pulling of a six-bottom plow.

43

Just as competent and hard-working in subfreezing temperatures as during the sweltering summer, this 460 Diesel was handy to have around for use with a front-end loader for snow removal. It was made from 1958 through 1963. The 460 Diesel—and the 560 Diesel—engine was a vast improvement over the previous IH diesel engines that started on gasoline and then switched to diesel. The 460 Diesel was a six-cylinder equipped with glow-plugs in the precombustion cup of each cylinder, which made cold weather starting much easier. Unfortunately, final drive failures tarnished the reputation of the 560 and 460 tractors.

REAPER WARS

The opening shot of what became known as the "reaper wars" was fired in 1843. Obed Hussey, a contemporary reaper manufacturer, challenged McCormick to a field contest between their two machines. McCormick and his machine emerged victorious. It was a single battle victory, but the war was far from over as it turned out.

Competition for sales was so fierce that various inventor/salesmen often devised innovative ways to capture the attention of farmers and their pocketbooks. One competitor performed such prodigious acts that he was called "The Reaper King." His chief feat that won him fame was to hitch himself in the place of a horse and pull one of his mowers to prove its light draft.

In a similar demonstration, another competitor, one of the Marsh brothers, bound an acre of grain in 55 minutes to demonstrate the simplicity of his harvester design.

The term "reaper wars" that's been tagged to this period was surprisingly accurate. Bare knuckle, face-smashing altercations between serious competitors were recorded during more than one field demonstration.

Patent litigation was costly and patent protection was somewhat limited anyway. As a result improvements in reaper and harvester design advanced rapidly due to designs being "borrowed" or licensed from competing manufacturers.

The letter series tractors easily became the IHC tractor story of the 1940s. The A, B, H, and M models lines, along with their counterparts in the McCormick-Deering line, were available with a wide range of standard and optional features. These included row-crop or standard-tread front ends; high-clearance models; orchard models; and gasoline, diesel, or in some models, LP gas engines. Power lifts and dual rear wheels were also some of the upgraded features offered to customers. Building on the successes of these models, IHC introduced the small and very popular Farmall Cub in 1947.

The 1950s was a troubling decade for IHC. Battles with unions stopped work regularly throughout the decade, including an eight-week strike at the end of 1958 that shut down work completely. Driveline failures on 460 and 560 models, introduced the same year, compounded the company's problems. These problems overshadowed the company's successes with the 100 series, featuring Farmall and International 100s, 200s, 300s, and 400s, released in 1954, and the Farmall 600 in 1956. The company also refused to adopt the three-point hitch, which was in widespread use throughout the industry.

THE 1960S AT IHC

As it neared the 1960s, the massive International Harvester Company was losing its grasp of the farm equipment manufacturing business.

To try to regain its engineering edge, IHC built the Farm Equipment Engineering and Research Center, called FEREC, on the IHC research farm at Hinsdale, Illinois. This facility consolidated engineering and fostered the sort of innovation that had begun to escape IHC's products.

"Behind [FEREC] was a phenomenal manufacturing and engine test facility," remembered Ralph Baumheckel, retired manager of International Harvester's Product Planning Research. "Granted the company had to bring casting in from some of the outlying foundries. Yet IHC could just literally manufacture everything for its farm equipment division right in that engineering center shop. Now for the first time all tractor engineering was located in one location at Hinsdale."

As for adequate research and engineering funding, Baumheckel commented that, "I can't totally answer that because when you sit on those committees you're prejudiced—you never get enough money. Deere & Company and IHC were the unquestioned leaders.

"Obviously Deere was pouring more money in than we were. But, especially in regard to the facilities, Deere was the only company that could even come close to us."

He noted that the big issues on the table in the tractor engineering area during the 1960s were "Hydrostatic transmissions versus Power-Shift transmissions versus alternate transmissions. This was high on the list, with IHC right up to the sale to Case.

"The objective was twofold. One was shift-on-the-go performance or infinitely variable performance, which was also shift-on-the-go or change-speed-on-the-go. IHC had looked at many different transmission configurations. The company had some

The Model 560 is able to cultivate four rows of corn at knee-high stage in early summer. Note that the "Farmall" designation has been carried on to yet another generation of tractors that stressed the tricycle-wheel design. The 560's new six-cylinder engine was built too strong for the "carryover" powertrain that was used. When problems surfaced with final drive failures, IH launched a massive, $19 million re-work program to assist dealers in correcting the problems.

Just 36 4300 four-wheel-drive tractors were produced between 1961 and 1965, including this 1961 model that was the sixth made. They were produced by IHC-owned Hough Manufacturing of Libertyville, Illinois. It has air brakes and an Allison auto transmission. Powered by a turbocharged 817-cubic inch engine developing 203-drawbar horsepower, the tractor weighed just 185 pounds shy of 15 tons. Too large for the farm machinery of the era, most ended up working in construction. Approximately only 10 of the units remain.

45

Why did a farmer make the necessary investment to take home a 4100 Four-Wheel-Drive? Those eight bottoms neatly turning over the stalk field soil quickly give the answer: One man and one tractor was able do the work of two men, two smaller tractors, and two plows.

outstanding engineering expertise in hydrostatics, so it ended up with the hydrostatic transmission.

"This actually started with an engineer coming on board right after World War II. He'd been involved in developing hydrostatics for anti-aircraft gun turrets. Hydrostatics had the advantage of being able to move from a creep to high speeds quite quickly, extremely precisely.

"And, you could stop it dead without any mechanical interface. This was done in the military, so when this engineer came to IHC he brought that expertise with him," Baumheckel continued.

"A hydrostatic transmission is a hydraulic pump infinitely variable that powers a hydraulic motor. In this case, it's a variable displacement pump driving a variable displacement motor. This can be accomplished in different ways, but this is how the industry has gone."

The tractors that were on the IHC menu when the 1960s opened were the latter editions of the Hundred Series. Introduced in 1961, the 404 and

Painted yellow instead of the more familiar IHC red, this 1964 2806 Industrial was typical: a heavy-duty front end, no three-point hitch, and no PTO. It was rated at 93 horsepower from a 301-cubic inch gasoline engine. Many units also mounted front-end loaders.

46

The biggest Wheatland model was the 1206D. Although still at 361 cubic inches, the efficient turbocharged diesel engine put out 112 horsepower. This unit was built in 1966. IH's first-over-100-horsepower two-wheel-drive tractor was the 1206 Turbo Diesel. The six-cylinder IH-built engine was basically a turbocharged 806 mated to a re-engineered driveline to withstand the increased horsepower output.

Built exactly in the middle of the 1965 to 1967 production run, this 1966 Farmall 1206D was the first farm tractor with a turbocharger marketed by IHC. The 361-cubic inch turbo diesel churned out 112-drawbar horsepower. The only engine repair work this particular tractor needed in more than 10,000 hours was a head gasket replacement!

47

504 IHC models were among those rolled out to counter the revolutionary Deere & Company New Generation of Power tractors. Improved Hundred Series members were "new" models introduced every two to three years that carried the designation 130, 240, 350, and so on.

Even doing two jobs, carrying two saddle tanks while spraying on an herbicide and injecting a fertilizer sidedress application, this 1966 1206 had power to spare. Operators also appreciated its platform height that kept them out of the dust more.

The noteworthy features of the Hundred Series tractors had already been mostly introduced in the later letter series tractors: Touch Control hydraulics, independent power take-off (IPTO), and Fast-Hitch (IHC's two-point hitch).

IHC opted to develop the Fast-Hitch as its own solution for mounting equipment. While it did offer an improvement to the sometimes erratic draft problems of Ferguson's three-point hitch system, it never captured the pocketbooks of farmers. Fast-Hitch was eventually replaced by what was becoming the industry's standard three-point hitch. Only IHC Fast-Hitch implements could be mounted on an IHC Fast-Hitch–equipped tractor.

The 404 and 504 reflected a change in farming that saw more customers desiring standard tread tractors over the row-crop tricycle versions. This was reflected in the fact that the IHC standard treads outsold the Farmall tricycles of the 404 and 504 models.

Other models in the line included tractors from the company's facilities in England, such as the McCormick International B-275. This utility tractor offered three-point hitch, four-speed transmission coupled with a hi-lo range giving eight forward and two reverse speeds, live hydraulics, and two-stage clutch.

In 1961, the B-414 diesel arrived from England, along with engineless B-414s that would take a gasoline powerplant. The gasoline model was then fitted with an engine made at the Louisville Works.

The "big" news for IHC and the industry in 1961, however, was the introduction of the 4300. At the time it was billed as the "World's Most Powerful Four-Wheel-Drive Agricultural Tractor." With its 300-horsepower engine, the 4300 was capable of handling a fully mounted 10-bottom plow.

The first prototype, called the 4WD-1, was judged to be underpowered. A decision was then made to boost the engine's horsepower into the 125 to 150 range. But when IHC personnel learned that the John Deere–articulated four-wheel-drive 8010 sported a 200-horsepower engine, they decided to up the ante. They'd built a bigger, more powerful tractor using IHC's DT-817 turbocharged diesel engine.

The 4300 was built by the Frank G. Hough Company, a subsidiary of IHC, located in Libertyville, Illinois. The units weren't an assembly line product, but instead were "stall-built" using many purchased components. The procedure also produced a rather high price tag.

The 4300 differed from the Deere & Company 8010 by being four-wheel-steer instead of an articulated configuration. They shared the same fate, however: Both were several years ahead of the times and so found few ready buyers.

IHC wasn't ready to quit on the "big" tractors, though. The company believed that the concept would eventually be viable. In 1965, it introduced the 4100. This tractor used many high volume production parts from IHC's own plants to reduce the cost.

Powered by an IHC-built turbocharged six-cylinder diesel engine, it delivered 116 horsepower at the drawbar. The following year it was also available with three-point hitch and 1,000-rpm PTO.

The 4300 and 4100 were painted yellow instead of the standard IHC red-based paint scheme. This was because the Hough facility was also building construction equipment, and that's the only color it painted. An upgrade modification, the International 4156, was produced during 1969 and 1970 as the next step in the company's four-wheel-drive evolution.

Although the 4300 was a "show stopper," what the line needed was some competitive models right around the century mark in horsepower. In 1963, the totally new 706 and 806 tractors moved into the marketplace to fill that niche. They proved themselves to be well-engineered, reliable units.

With the 560 problems still fresh in their minds, the engineers took no shortcuts with these two tractors. The 806 engine was entirely new and designed especially for the new model. The 706 engine had

All original, this TB25 Series B has just 351 hours on its tachometer and was never outside until 29 years after it came off the assembly line in 1966. The reason: The tractor was used at a U.S. government technical training center. The 265-horsepower engine gives the crawler enough oomph to pull a 14-bottom plow.

49

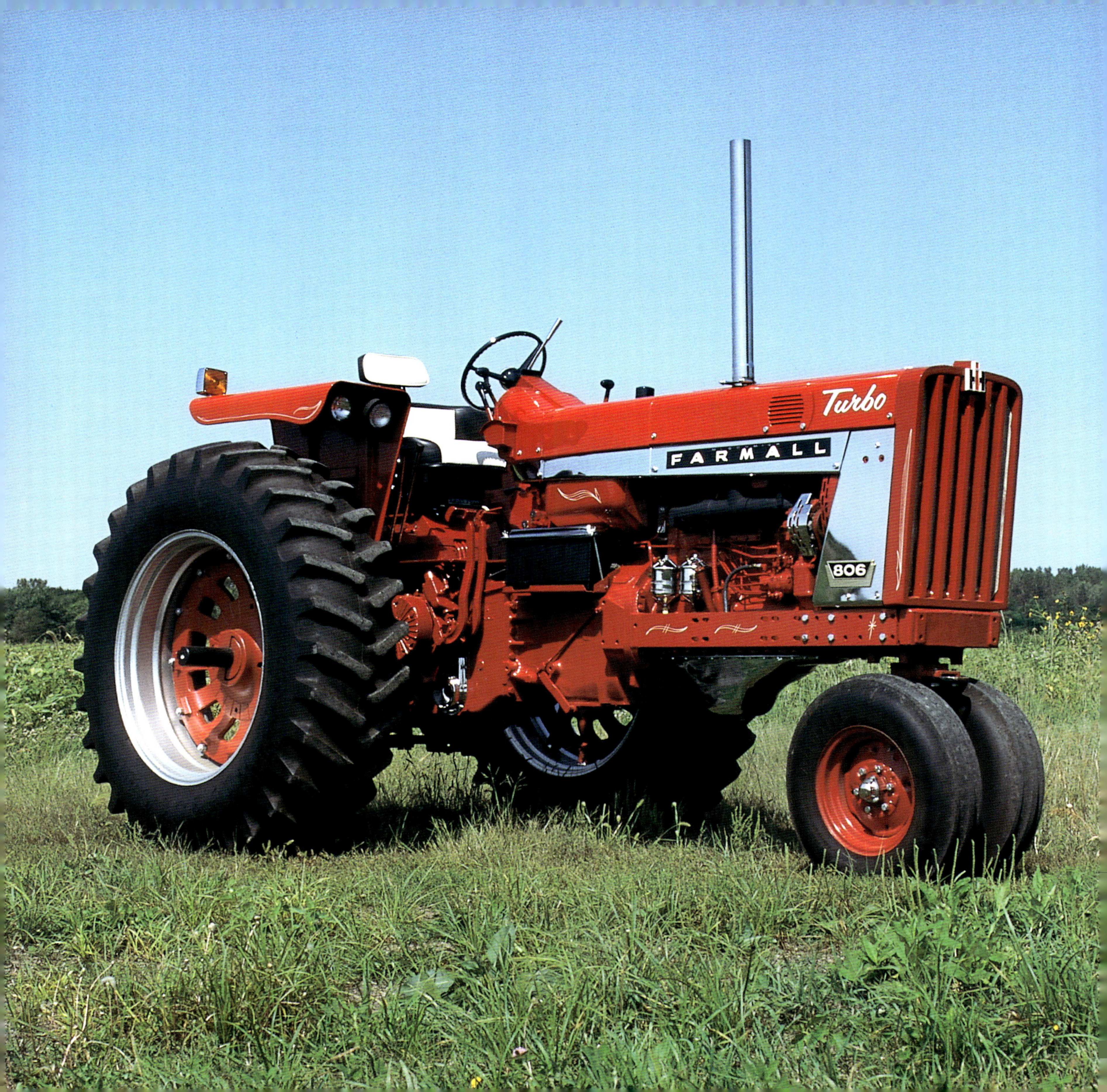
Turbo
FARMALL
806

This 1967 Farmall 806D has been "chromed" to make an even more eye-catching parade tractor. The series was made between 1963 and 1967. The 94-horsepower diesel engine had 361 cubic inches, plus an added turbocharger that ran at 145 rpm.

been previously used in other IHC products and had proved itself reliable. Besides the popular diesel engines, these models were also available in gasoline and LP gas versions.

The transmission, final drives, and axles were all designed from the ground up to complement the higher horsepower of the engines. And, just to make sure IHC had it right, one tractor endured 75,000 hours of on-the-farm testing to work out the bugs before a customer ever got his hands on a production model.

Refinements included a deluxe seat positioned ahead of the rear axle for a smoother ride and an uncluttered platform for the operator. The PTO was a new independent design featuring both 540-rpm and 1,000-rpm shafts so no changeover was necessary regardless of which speed was needed.

The Power-Shift torque amplifier was a new design for these tractors. It proved so reliable that it was used in later, larger horsepower tractors too.

The 706 and 806 three-point hitch was designed around the torsion bar, lower-link sensing system. Front-wheel assist was a factory-installed option, and the rear axle could accommodate dual tires. Both of these features helped capture more of the horsepower these tractors churned out.

The Farmall 1206 Turbo Diesel was basically an 806 on steroids. By turbocharging the engine and strengthening the driveline, the 806 was transformed into IHC's first two-wheel-drive "more than 100 horsepower" tractor. The 1206 entered the IHC agricultural tractor line in 1965.

As the 1960s drew to a close, several issues plagued the company. In the early part of the decade, the company began to upgrade its dealerships and pare down the number in the network until it had only 4,500 full-line dealers; but a lot of these dealers weren't entirely happy due to the heavy-handed manner the company exhibited in negotiations.

Add to this a construction equipment line that was considered by some to be more liability than asset. The company's poor reputation for quality was

This tractor had all the power needed, speed, and enough agility to run down the rows without damaging the crop, soybeans in this instance. Although designed for heavier pulling jobs, the bigger IHC tractors like this one also had the flexibility to do most types of fieldwork.

NAME RATIONALE

Ralph Baumheckel, retired manager of International Harvester's Product Planning Research, had this to say about the dizzying array of model names and the justification behind building two tractors that were either identical or similar, selling one as an International, one as a McCormick, one as a Farmall, and some with a combination of those names:

"There is a rationale behind it. Let's go back to McCormick and Deering. Let me bring you up to the 1960s. I've been told that decision was made in a naming committee. When I asked this question, I was told that historically certain parts of the country had been dominated by Deering products.

"So when we were marketing to that area we wanted the Deering name to be there. Other areas had been dominated by McCormick, and so this was strongly McCormick sales territory. That's why the McCormick and Deering names got switched around so much. It was strictly to take advantage of what was perceived as a marketing advantage," he said.

"And this gets even more confusing too because the Farmall was first known as a tricycle tractor; however, in the 1930s it was also available with a wide front axle. This related more to clearance and the way it was spaced out. So when the tricycle tractor began to give way to the wide front end we moved to emphasizing the International name.

"It was interesting in that in certain configurations you would see this big International nameplate on the side of the tractor, and then down underneath the model number in small letters would be Farmall," he added.

"Farmall had been the name that had made International. No wonder the company was hesitant to let 'Farmall' totally disappear from its tractors. So if it had the name Farmall, it was more of a farm tractor in terms of transmission speed and clearance, even with a wider front axle.

"If it was International, period, the Internationals were on smaller tires, set lower to the ground, and traditionally had fixed, nonadjustable rear axles. A farmer was able to slide the rear wheels in and out on a Farmall's rear axle, however.

"Consequently, that tractor was still called a Farmall. If you lost the ability to shift the rear axles, then the tractor was a classic International. In other words what they coined 'utility' was an International with a little different model number," Baumheckel continued.

"And then we confused it even more. We added an industrial line that was given a model number by putting a 2 in front of the usual model number. This tractor would be identical to a farm tractor, but it would be painted yellow and sold by the industrial equipment dealers."

Designed for the larger equipment starting to come out, the Farmall 1256 made between 1967 and 1969 put out a solid 116 horsepower from a 407-cubic inch turbocharged diesel engine. This is a 1968 model.

52

also impacting sales. The crawlers and industrial tractors suffered some of the same problems that bothered the 460 and 560.

THE AFTERMATH OF GREATNESS

From Barbara Marsh's, *A Corporate Tragedy*, this observation reflects the extent of erosion of IHC's position as a major entity:

" 'I wish the new management team a lot of success,' a retired Harvester executive said of Menk, Lennox, and their executive cadre in 1982, 'but, I don't know them and I don't know the company they work for.' The Harvester this man remembers, the place where he spent all of his working life, ranked 27th on the Fortune 500 list with sales of $8.4 billion in 1979. It boasted of profits of almost $370 million.

Demands of the 1960s vastly increased the need for larger equipment to cover more acres with less labor. This, in turn, sparked the requirement for bigger-engined tractors. At the beginning of the decade, before such tractors were commonly available, a few innovative farmers hitched up two tractors in an attempt to double horsepower.

Relying on only a hand clutch and with no three-point hitch and no PTO, this 1968 2856 Industrial also developed 100 horsepower, although its diesel engine had 407 cubic inches.

One of the first tractors to be equipped with an IHC cab, this is a 1969 756D Wheatland. Its 310-cubic inch diesel engine developed 76 horsepower. When the 56 Series was introduced in 1967, IHC factory-installed protective frames were made available to comply with Federal safety regulations.

A high clearance model that provided more daylight between tractor and growing crop, this 1968 Farmall 856 developed 100-PTO horsepower from a 407-cubic inch six-cylinder IH-built engine.

"It employed nearly 98,000 people around the globe, including 64,000 in the United States, and operated 41 manufacturing plants worldwide. It stood as the leading North American producer of heavy- and medium-duty trucks, the world's No. 2 maker of farm equipment, a major presence in construction equipment, and a top manufacturer of gas turbines.

"By 1983, a drastically shrunken Harvester had sales of $3.6 billion, placing it 104th on the Fortune 500. It ran up losses of $485 million that year, making a total of nearly $3 billion in losses for 1980–83. It employed 32,000 workers worldwide, including 19,000 in the United States. It had trimmed its overextended industrial empire back to three businesses, farm equipment, trucks, and engines, and it operated just 27 plants around the world."

In 1984, in the midst of the severe 1980s agricultural depression, recession was just too mild a description. IHC sold its farm equipment division assets in North America and Europe to the J. I. Case Company, which was financed by Tenneco Inc. of Houston, Texas.

Case IH continues to incorporate some of IHC's 153-year legacy of innovative farm equipment manufacturing under this new structure.

Powered by a diesel engine rated at 95 horsepower, the 826 was a popular model. Now, more than three decades after being made, this 1969 model is still doing fieldwork. First offered in 1969, the 826 could be ordered with a conventional gear-drive transmission or an infinitely variable hydrostatic drive. Standard features were power steering, three-point hitch for Category II implements, and hydraulic power brakes.

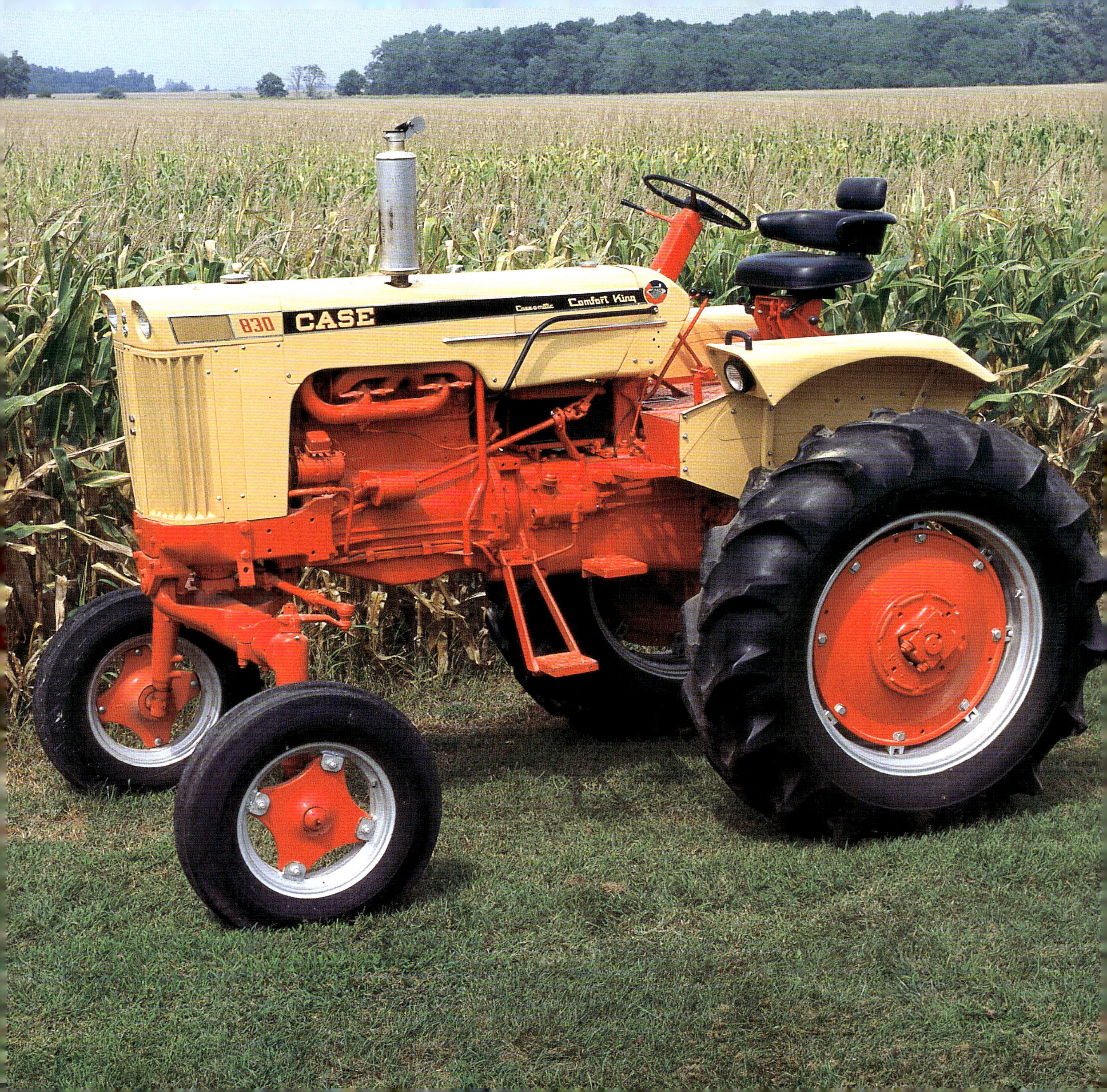
830
CASE
Case-o-matic
Comfort King

CHAPTER THREE

J. I. CASE COMPANY

Jerome Increase Case was the third member of the nineteenth century pioneering triumvirate that revolutionized the world's food production. John Deere provided the basic tool for seedbed preparation—the steel plow. Cyrus McCormick followed with a mechanical advancement in harvesting grain—the reaper. And it was J. I. Case who developed the thresher into a fast, labor-saving, and economical way to separate the grain from the head.

Case was born December 11, 1819, in Williamstown, New York. In 1842, when Case was 22 years old, he left his home just northeast of Rochester, New York, and set out for Rochester, Wisconsin. He brought with him six new threshers that he had purchased on credit. He'd sold five of them by the time he reached Rochester, where he spent two years improving the sixth.

In 1844, he demonstrated his improved machine, which combined the threshing and cleaning operation in one unit. When he could not get hydraulic water power for a manufacturing operation in Rochester, he moved to open a shop in Lake Michigan's new power town, Racine. Six threshers were immediately built and sold in time for harvest, and more orders were on the waiting list. In just three years Case became the largest employer in Racine.

Growth was swift, and by 1900, the J. I. Case Threshing Machine Company entered the new century well positioned in a field crowded with farm equipment manufacturers. Its 1900 catalog stated, "We are the greatest manufacturers in the world of engines and threshing machinery." While threshers would run their course by the 1930s, by that time, Case was a prominent manufacturer of tractors.

Case worked with a gasoline tractor engine in 1894, but abandoned the effort by 1897, in favor of its steam-powered machines. By 1910, Case would come full circle, entering the gasoline tractor market with its Model 60. The company never looked back.

For 15 years, 1914 to 1929, Case tractors had the distinctive crossmotor design, with the engine mounted across the tractor frame instead of parallel. The transverse, or "X," engine looked unique but sold well enough that Case began exporting them to several foreign countries.

Case tractors offered in the 1930s and 1940s could justly be called the "alphabet line." Customers could choose from the C, CC, CO, CI, D, DO, DC, L, LI, LA, LAI, R, RC, RO, S, SC, SI, SO, V, VA, VI, VAI, VAC, VC, VAO, VAH, or the VAS models. This line evolved with the addition of numbers, such as 1, 2, 3, and 4, to become the DC-3, DC-4, CC-3, CC-4, VAC-11, VAO-14, and VAC-15, to name a few. The "heavy iron" Model L was a best-seller during this period, racking up some 30,000 sales.

In the immediate postwar period through the mid-1950s, Case had a conservative, bare-bones line of tractors and equipment. These models had been great in the 1930s, but fell short of the needs and demands of postwar farm practices.

To compete with the Ferguson three-point hitch, Case introduced its Eagle Hitch on the VAC model in 1949. Company personnel felt its design of draft control was vastly superior to the Ferguson system. The Eagle Hitch, in various forms, was used until 1964.

Even though the tractor division was behind the curve in comparison to its competition, 40 years of research, design, testing, and learning from the industry

Both big and tall, the 1964 830 Case-O-Matic Comfort King Row-Crop Hi-Crop was designed specifically for work in the sugarcane fields in Louisiana and Florida. It mounts a diesel engine of 65 horsepower.

A small, 126-cubic inch gasoline engine powered this row-crop with wide front-end 1960 Case 200 Tripl-Range. Manufactured mostly in the 1950s, the B Series was offered in 200, 300, 400, 500, 600, 700, 800, and 900 models. The 400, 600, and 800s came with Case-O-Matic, while the others were made only with straight clutches.

A matched pair of 800 Diesel tractors, each equipped with Case-O-Matic, carry out the last cultivation of growing corn for the season. The Racine Plant tractor production log lists only 145 800s produced in 1958 and 110 made in 1959. Obviously it wasn't a high production model. Case's four-cylinder engine gave the 800 Diesel 54-PTO horsepower.

It was quite unusual for a 1962 630D Orchard tractor to have both Case-O-Matic drive and a hand clutch, as this one does. The diesel engine puts out 48 horsepower. The tractor provides almost the ultimate in protection while doing tree work.

59

In addition to Case-O-Matic drive, this 1962 630D Orchard tractor also gives the driver the option of using the hand clutch on the left side. With the introduction of the 30 Series in 1960, the line included an expanded line of grove and orchard tractors with full grove sheet metal. The grove and orchard tractors were drawbar models only.

had nevertheless brought many refinements to Case tractors. PTOs, hydraulics, electric starters, lighting packages, and seat refinements all contributed to a more user-friendly tractor, but farmers still wanted more.

In 1953, Case released its 56-horsepower diesel Model 500. More new models followed the 500 as Case worked to upgrade its tractor line. The 300 was an upgrade of the VA from its Rock Island facility. The standard transmission was four-speed, but customers could opt for the new 12-speed Triple Range-transmission. The Model D line gave way to the all new Model 400 series in 1955.

Case's late entry into the crawler business came via its purchase of the American Tractor Company (ATC) of Churubusco, Indiana, in 1957. The ATC line of crawlers carried the name TerraTrac, renamed the Case crawler after the merger.

Marc Rojtman, the president of ATC, became president of Case in 1958. Rojtman thrust the company into the modern era of manufacturing and marketing, stirring up the lethargic board of directors and the ultra-conservative administration. He expanded the marketing network at home and abroad, expanded and expedited the product line, and pushed for an expanded materials handling line that would ultimately lead to the Construction Equipment (CE) division becoming a major part of the company.

His new marketing program raised an intense amount of interest in the new products; however, many dealers and farmers simply didn't have the financial means to take home what they saw and liked. In response, Rojtman established the Case Credit Corporation, putting more Case tractors on farms, more construction equipment on job sites, and a lot more receivables on the books.

The Case-O-Matic transmission debuted at the 1957 World Premiere. Rushed to market, it had a few weaknesses. Though the company quickly corrected them, a certain hesitation over the transmission lingered in the public mind.

CASE'S OTHER ENDEAVORS

When J. I. Case took in three partners in 1863, he made it quite clear that they'd be expected to devote full time to the company business since he wanted time to pursue other interests.

Case helped found the Wisconsin Academy of Science, Arts, and Letters. He also served as president of the Racine County Agricultural Society and the Wisconsin State Agricultural Society. Three times he was elected to the Wisconsin Senate, and he served as mayor of Racine for two terms.

Case was a founder and later president of Racine's Manufacturers' National Bank and the First National Bank of Burlington, Wisconsin. He was instrumental in helping establish new banks in several other states too.

He also invested in several cargo ships that for decades plied the waters of the Great Lakes, carrying grain to Buffalo, New York, Chicago, and other ports.

Good horseflesh appears to be the one indulgence that was entirely separate from Case's business interests. His Hickory Grove farm just south of Racine on Lake Shore Road was home to record-setting pacers and trotters. His famous black trotter, named Jay Eye See (as in J.I.C.), even became the subject of a Currier and Ives engraving.

The farm and stable of fine horses were sold by his son, Jackson Case, after Case died.

The 1962 430 LCK, "Low Center King," Tripl-Range sports wheels and tires that make it ideal for golf course maintenance or similar work around the farm. It has 34 horsepower. Note the adjacent famous and also restored Case eagle.

Another development that year was a new look for Case tractors. The round nose was made more square, with recessed headlights above the grille. In addition, a black stripe was added on each side of the hood, a touch that became a Case hallmark until the 1990s.

CASE IN THE 1960S

In February 1960, the Rojtman era came to a close and William Grede, founder of the Grede Foundries and a Case board member since 1953, replaced Rojtman as president. Grede inherited all the promises and all the problems of the company.

James Ketelsen became assistant controller at Case in 1958, and later assumed the role of treasurer and vice president of finance. He advanced to the company presidency in 1967.

Here's how Ketelsen viewed the 1960s: "We entered that decade of the 1960s in bad shape. It was traumatic. Rojtman had been the driving force in the late 1950s expansion. He woke the company up. Yet there were pluses and minuses there, of that there was no question.

"It was his management that really pushed aggressively on developing and expanding distribution, expanding market share, and crashing through engineering programs of product.

"And some of those engineering and development programs were too rapid, not well tested, and

Row-Crop 730 with 58 horsepower is the high-profile model, which means its hood is raised higher than normal. This allows the fuel tank capacity to be increased from 22 gallons to 32 gallons. The grille is also larger. Most similar 1963 models lacked the wide front end of this particular unit.

61

Mutt and Jeff, David and Goliath: The 1964 130 compact tractor that could almost be driven underneath the tall 1964 830 Case-O-Matic Comfort King Row-Crop Hi-Crop tractor shows the great contrast in models of this era.

They might have "Garden" as a first name, but Case sold thousands of these small, "compact" tractors after buying the line from Colt Manufacturing in 1963. Twenty years later, Case sold this line to Ingersol Tractors. From left, they're a 1966 150, 1964 130, 1964 180, and a 1967 195. The 150 was a second-generation garden tractor with 10 horsepower. The 130 also provided 10 horsepower. The 180 was Case's first 12-horsepower garden tractor. It was followed by the third-generation 12-horsepower 195.

62

Although less sheet metal was added to the orchard model than in earlier years, this 1965 530 Utility has the "grove fenders" virtually necessary for working among fruit trees. It channels 41 horsepower from its diesel engine through a Tripl-Range drive that provides 12 speeds and road speeds up to 25 miles an hour.

then they put a bunch of product out on the market, both in the construction and farm equipment end, that had a lot of trouble," Ketelsen added.

"This coincided with this excess inventory position so the company had too much inventory out in the field that hadn't been sold and that also had engineering problems. It was a tough situation.

"What happened was that the company took back a lot of product, re-engineered, and then got good product out into the marketplace. In addition it cleaned up the dealers' excess inventory situation. So by 1964 the company was, from the business standpoint, in pretty good shape. Case still had a weak balance sheet.

"Then Kern County Land Company from California acquired a majority interest and added needed capital. That solved the balance sheet problems as well. From then on Case began to do better," he said.

In an effort to reduce costs after Rojtman's departure, all tractor production was consolidated at the Racine facilities. This meant that the 430, 530, and 630 were produced at two locations through a transitional period.

The 430 was rated at 35 horsepower. It included as standard equipment a four-speed transmission. The 12-speed Tripl-Range transmission was available as an option, as well as the eight forward and reverse

The 1200, this one a 1966, was the only four-wheel drive in which the front wheels were steered with the steering wheel and the rear wheels with a hand lever or foot control. What it could do: move kind of sideways while making four tracks in the dirt—called "crab steering." A 24-volt electrical system was used. Its 451-cubic inch turbocharged six-cylinder diesel was built by Case itself. The transmission provided eight forward and four reverse speeds.

63

You won't see this much heavy sheet metal on the fenders or cowling of a large orchard or grove tractor today, because most such trees are kept more closely trimmed nowadays. The 1967 Dual-Range 730 Orchard and Grove's diesel engine develops 58 horsepower.

speed Dual-Range shuttle transmission. The Snap-Lock Eagle Hitch system made hooking up mounted equipment much easier than conventional three-point systems.

A diesel was available in a 188.4-cubic inch, four-cylinder engine for the 430. Both gasoline and diesel engines were offered in tricycle and standard-tread models.

The 530 upped the rated horsepower by 6 over the 430 and featured the Case-O-Matic drive. It offered all the options of the 430 except a diesel engine that didn't become available for the 530 until later. An LP gas engine was an option on the gasoline model.

The 630 had a production run of three years that ended in 1963. It came in six models: dual front wheel, single front wheel, adjustable front axle, standard four-wheel, Western Special, and Orchard Special.

The 630 diesel was the first Case diesel engine to use direct injection as opposed to the Powrcel combustion system. Standard transmission was the 12-speed Tripl-Range. The 630C used the Case-O-Matic drive.

From 1960 to 1969, the 730 offered five-plow power for big wheat farming and row-crop operations. These tractors became part of the Comfort King series from 1965 until 1969. The 730 models offered three fuel options: gasoline, diesel, and LP gas. Drawbar horsepower was at the 50 level from a four-cylinder Case-built engine.

The next step up in the 30 series was the 930. This tractor introduced the option of a 1,000-rpm

"The problem comes in when these tractors get higher and higher in horsepower, yet the farmer wants his tractor to run at least until noon before having to refuel," he added.

"We wanted to have an 80-gallon fuel tank. Well, if you can visualize a 55-gallon drum, where do you put it and hide it? So our solution was to go to a rear-mounted fuel tank. With a plastic fuel tank installed we rolled a tractor over as many as eight times without rupturing the fuel tank."

In 1966, the Comfort King Model 1030 diesel was Case's new tractor in the 100-horsepower range. It exclusively used the six-cylinder, 451-cubic inch diesel engine designed and built by Case. Rojtman was gone, but some of his legacy was still felt by the tractor engineering division.

When Grede assumed the presidency in 1960 and tried to throttle down the Case express to bankruptcy, he immediately pared engineering staff and budget. The Case engineers' dilemma paralleled the often-told story about employees doing so much with so little for so long that they could do anything with nothing.

In the case of the 1030, it became apparent they'd tried to do too much with too little in regard to the transmission. The six-cylinder engine was

The air cleaner of the 1967 730 Orchard and Grove is carried in a low position and designed not to catch tree branches. The exhaust also is directed down and to the side of the tractor. Transmission options were either the Case-O-Matic or dry clutch on the 730 Orchard and Grove models.

PTO. This was a bigger five-plow tractor at 66 horsepower, available in standard tread and as a Wheatland or Western Special model.

In 1964, power was boosted on the 930 by marrying a six-cylinder engine to the 730/830 chassis. Now Case finally had a general-purpose and row-crop tractor in the 85-horsepower class to compete in the escalating horsepower race. The Draft-O-Matic hitching system also debuted the same year on the larger tractors.

Eldon M. Brumbaugh, who headed the engineering department at the tractor engineering facility in 1956, explained some other features offered on the 930: "Case was the first company with the isolated platforms on tractors. On the 930 series, we were also first with the rear mounted fuel tank and plastic fuel tank that most everybody uses today.

THE "OTHER" J.I. CASE COMPANY

In 1876, Case financed a new company, independent of the Threshing Machine Company, which became the J. I. Case Plow Works. The senior Case removed himself from management of the Plow Works in 1880 and named his son, Jackson I. Case, as president.

Soon the family began squabbling. Legal action resulted in the Wisconsin Supreme Court handing down a 1915 ruling that said the Case Threshing Machine Company could make farm implements, particularly plows, as long as the name Case or J. I. Case didn't appear on the equipment.

Case Threshing Machine Company purchased the Grand Detour Plow Company in 1919, providing it with a source of plows, cultivators, and listers without having to deal with the "other" firm across town.

The tale of two Case companies was resolved permanently in 1928 when the Plow Works became an acquisition of Massey-Harris. There was one final development, though.

Massey-Harris wanted the product line and the distribution network of Case Plow Works, but it didn't want, or need, the Case name. As a result, the Case name was sold to the J. I. Case Threshing Machine Company for $700,000. From that point on there was only one Case firm.

About this time the company decided to change its name since there was again only one Case company and the threshing machine business was no longer strong enough to justify the words threshing machine in the company name. In 1928, the firm's official name was changed to J. I. Case Company Inc.

65

available and adequate; however, a suitable transmission wasn't. So the engineers stretched the 930's tranny in hopes of providing the bigger row-crop tractor that the market demanded. It was less than a perfect union as reliability problems developed after sales were under way.

The 1030 was produced from 1966 to 1969 with production numbers of 13,763. Even with its problems, it was second in Case tractor production during those years only to the 19,164 930 units produced.

Early in the 1960s both IHC and John Deere broke the 200-horsepower level. IHC sold a 214-horsepower 4300 model four-wheel-drive tractor, and Deere & Company marketed an articulated four-wheel-drive 8010 rated at 215 horsepower.

Case responded in 1964 to the call for bigger power with its first four-wheel-drive tractor, the Model 1200, which also featured four-wheel steering. This turbocharged 451-cubic inch diesel model was named the Case Traction King Diesel. Unfortunately, the engine could only muster 115 horsepower. This wasn't nearly enough to satisfy marketing personnel, who would have to sell it against the competitions' tractors in the same price range.

Case just didn't have another engine option available. So engineering "tweaked" the naturally aspirated unit by adding a turbocharger. The Lanova diesel design proved a challenge to turbocharge without overheating the combustion chamber to dangerously high temperatures.

A work-around solution was the mounting of a pyrometer on the dash to warn the operator when temperatures were reaching the danger level.

The 1200 could rightfully be called a "shoestring" project. Regardless, it overcame many difficulties and established Case as a major contender in the big four-wheel-drive market.

Case was also expanding its horizons in other areas of the company during this time. The construction equipment (CE) division offered a wider line with Case-designed and -built loaders and backhoes, plus an all-terrain forklift utilizing the 430 chassis. Outsourced items such as cranes, log forks, and dozer blades also became part of the CE line.

As many of its competitors were doing, Case entered the lawn mower field in 1964 with the acquisition of the Colt Manufacturing Company of Winneconne, Wisconsin.

A milestone of the 1960s for the Case line was the introduction of the company's first Power-Shift transmission in 1969. It first appeared on the 870 Agri-King Power-Shift Diesel.

Perhaps the milestone with the most far-reaching impact for Case came in 1967 when the Kern County Land Company was acquired by Tenneco of Houston, Texas.

The 1967 930 Western Special Wheatland is unique in having a six-speed transmission and a hand clutch. Other 930 row-crop models had an eight-speed transmission, foot clutch, and Draft-O-Matic. Its diesel engine provides 85 horsepower.

WAR PRODUCTION

Like other agricultural equipment manufacturers, a significant part of the Case manufacturing capacity was shifted to war materiel production during World War II. Case made components for Army trucks, Sherman tanks, and the Navy's "Water Buffalo." Other war materiel contracts included aftercoolers for Rolls-Royce engines used in the P-51 Mustang fighter and the Mosquito bomber, B-26 bomber wings, .50-caliber gun barrels, and 155-millimeter shells.

Specialized tractors constituted a large part of Case's contribution to the war effort. These were lightweight tractors that could be dropped via parachute, armored tractors, tugs for aircraft carriers, high-speed models that could move 40 miles per hour, and others. More than 15,000 Case tractors joined the military.

Everything, including paint, is still original on this 1967 930 Comfort King Row-Crop. The six-cylinder engine has 377 cubic inches. Operators liked it, because with the LP gas tank behind the seat there was no tank to obstruct the view or take up room under the hood.

Three things set this 1968 1030 Comfort King Western Special apart from the Row-Crop model: It carries the big fenders, the front end is set back toward the rear a little farther, and there's no three-point hitch at the rear. The 451-cubic inch engine puts out 101 horsepower.

AGRICULTURAL TRACTORS TRAIL CE

As the 1960s drew to a close, the passing decade proved to have had both good and bad years for the 126-year-old Case company. Having begun the 1960s with "bankruptcy at the door" and deep in debt, Case shifted focus to construction equipment and emerged a strong contender for that market. In 1969, CE sales topped the ag division for the first time in company history.

Michael Holmes said in *J. I. Case: The First 150 Years*, "Between 1969 and 1971, Case's entire construction equipment line had been replaced. In 1971 alone, the company introduced more new machinery than any of its competitors. It had also become the nation's third largest producer of construction equipment.

"Retail 1971 CE sales were up 36 percent over 1970. That achievement, plus a turnaround in the farm economy, helped Case lead all Tenneco companies in earnings gains for 1972."

With a 451-cubic inch turbocharged engine that provided 101 horsepower, the 1030 could be used for just about any field task. This particular tractor mounts one of the early cabs designed to give the driver some rollover protection while also keeping him more dust-free and comfortable.

Unusual for a Row-Crop model, this 1969 770 Agri-King also sported Western style fenders. It was the smallest of four models introduced at that time: 770, 870, 970, and 1070. The 770's diesel engine developed 56 horsepower.

69

This was a fairly typical scene during the springtime on many Midwestern farms: Case tractor effortlessly getting the acres covered with a minimum of fuss.

NEW ERA OF MARKETING

In December 1957, Case set a precedent for introducing a new agricultural equipment line. Before this event, Case, as well as other manufacturers, introduced their new models at regional shows.

Case's Rojtman, who came with the ATC acquisition, was the epitome of the promoter personality. He envisioned a much bolder and flashy marketing program. Thus a Case "World Premiere" was planned for Phoenix, Arizona.

This was the first time in history that an agricultural equipment manufacturer brought all of its dealers to one location to view a new line of products.

For six weeks a rotation of Case dealers, their wives, finance people, and others in the industry were flown in. During their three-day visits they examined the new products and participated in the festivities. Rojtman even invited several hundred competing equipment dealers.

More than 3,000 attendees viewed the display of Case products. Featured were the new Case-O-Matic transmission and other items that were heralded as "Case tractors years ahead in styling and performance." Showcased were 12 distinct power sizes and 124 models.

Garnering much attention in the center of a circus tent was a tug-of-war staged between the Case-O-Matic and a competitor's tractor of equal horsepower. The tractors were hooked back-to-back, drawbar-to-drawbar. The competitor's tractor initially began to pull the Case backward, to the dismay and disappointment of the crowd.

But once the torque converter multiplied engine torque, the Case tractor started to pull the competitor's tractor backward. The crowd loved it, as did the dealers. This bit of showmanship helped rack up 30,000 tractor sales during the course of the World Premiere.

This prompted *Tooling and Production* magazine to list Case as one of the best-managed U.S. corporations in 1967.

A NEW CASE OF A DIFFERENT COLOR

It never hurts having a friend in high places to champion your cause. Case's such friend was James Ketelsen. He, more than anyone else, is credited with the "game-saving" financial plan for the company in 1962 when Case couldn't meet $145 million of short-term notes.

Ketelsen later went on to join the Tenneco corporate staff in 1972. There he championed the Case cause when others at Tenneco felt the company should be sold. Case certainly needed a friend like Ketelsen during the farm recession that started in the early 1980s and blossomed into an agricultural crisis.

A critical decision was made about the company's future in 1984. The company was, as Ketelson described it, "too small to challenge Deere and Company, too big to abandon, and too unprofitable to sell." Ketelsen, working with then Case President

Jerome K. Green convinced Tenneco to hang on to Case and explore some options.

The main "option" was International Harvester Company. IHC realized in 1984 that it, too, needed to explore some other options. Case, with Tenneco dollars, proved to be the solution.

The package called for roughly $475 million to make IHC's agricultural line, production facilities, and distribution system part of Case. The effective date for the deal was early 1985. By a year later the two companies had completed most of the necessary integration.

The two combined companies again gave Case a full line of products with the exception of hay and forage equipment. IHC had previously sold its hay and forage line to New Idea. So Case formed a joint venture with Hesston for this company to provide these products. Later Case acquired the Steiger Tractor Company of Fargo, North Dakota.

The new company brand name became "Case International" with the "Case IH" badge. Also new was the color scheme of Harvester Red sheet metal and Case Black chassis and decals.

The legal name of the company remained J. I. Case Company until 1994. Then the corporate name was changed to the Case Corporation and a new logo, or badge, emerged.

With the added IHC dealerships and Tenneco's financial resources, Case's Harvester Red was positioned second only to the other "color" in the agricultural equipment game, Deere and Company green.

Case posted 1997 revenues of $6 billion and sold its products in 150 countries. Its marketing network consisted of approximately 4,900 independent dealers worldwide.

Jerome Increase Case could surely have related to both the trials and triumphs his company encountered over the years. Because, he, too, endured many of both during his lifetime.

True, Case was bought by the Kern County Land Company, which was in turn bought by Tenneco, and then joined with International Harvester Company to become today's Case IH.

This has simply been the continuation of the cycle of events that have occurred in the agricultural implement business since the beginning of the industry. Growth and leadership have come from the joining of ideas, people, and facilities, both at home and abroad.

Ketelsen's present assessment of Case is, "Case IH is excellent now. In the end this has worked out well. I think Case IH is a good, strong company, a good competitor to Deere and Company and one that gives the farmer a choice in terms of equipment.

"Overall, all this probably keeps prices more favorable toward the consumer than they'd otherwise be."

The first model of this four-wheel drive came out in the last year of the decade. It was a follow-up to earlier four-wheel-drive models, such as the Model 1200, and ideal for wide expanses of land unencumbered with fences or terraces.

71

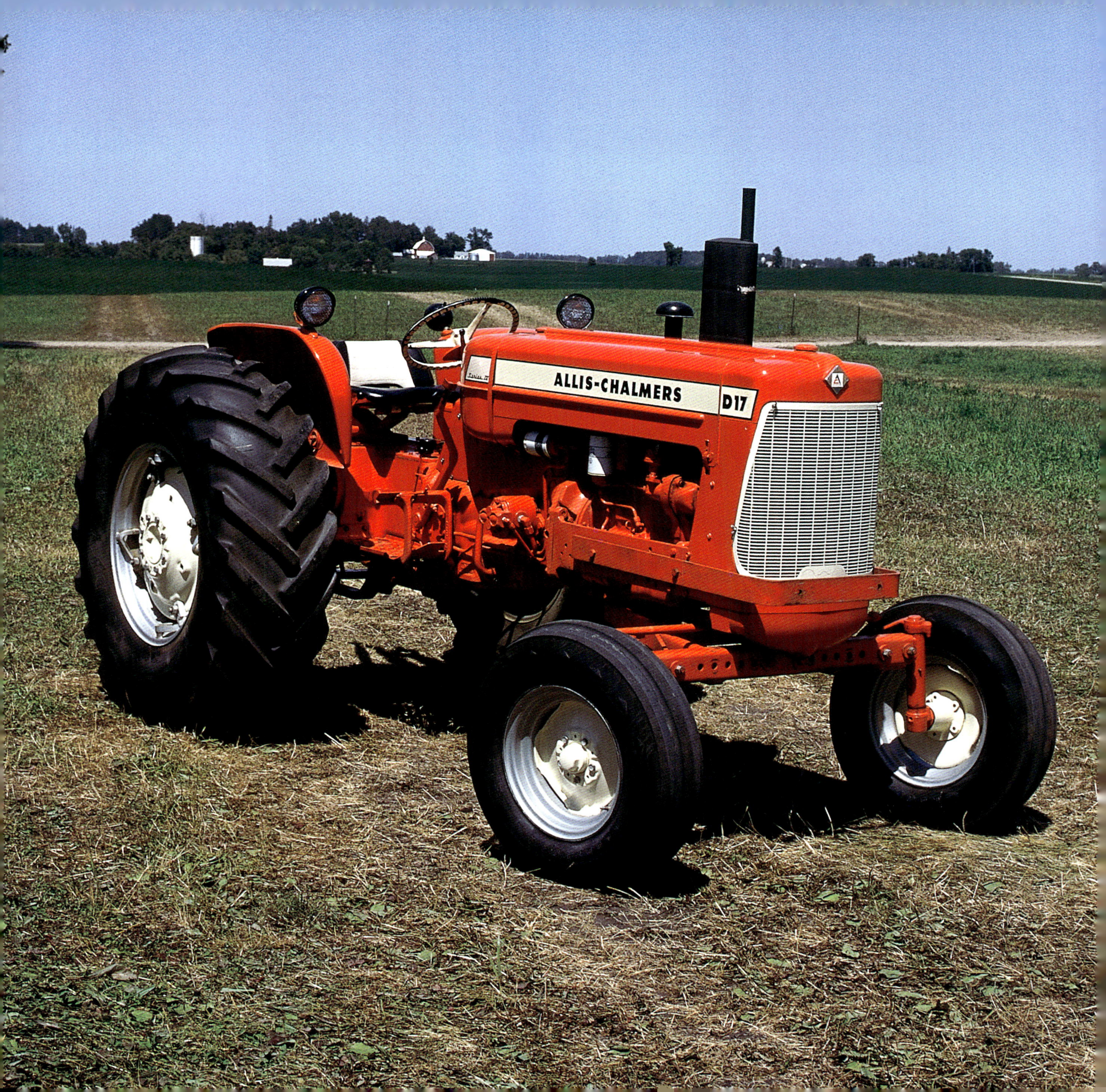
ALLIS-CHALMERS
D17

CHAPTER FOUR

ALLIS-CHALMERS MANUFACTURING COMPANY

The roots of the Allis-Chalmers story include three principal individuals: Charles Decker, James Seville, and Edward P. Allis. Decker and Seville moved from Ohio to Milwaukee, Wisconsin, in 1847. Allis, a New York native, found his way west to Milwaukee sometime before 1846. The first two men built and grew the company of "Decker and Seville—Manufacturers of French Burr Millstones, Grist, and Sawmill Supplies." But they grew it too fast and too big. In 1861, Allis picked up the whole business at a sheriff's sale for $22.72. He reopened it as Edward P. Allis and Company.

Allis excelled at two areas of management: He seemed to know what his customers wanted to buy. So the product line grew quickly to satisfy their needs. And he let his key people do their jobs without undue interference from management.

Sales and income climbed to $100,000 in 1865, $150,000 in 1869, and $350,000 by 1870. By then, the work force had ballooned to 200 people. The company quickly outgrew its Reliance Works facilities. A new plant site was chosen at Clinton and Florida streets. Construction began in 1866, and the building was occupied in 1867.

In 1869 or 1870, Allis purchased Bay State Iron Works to give his operation some much-needed additional foundry capacity. Allis died in 1889, but the company continued with other family members in key roles. It expanded again, buying land in what would become West Allis, Wisconsin, in 1900.

The company entered several new fields of manufacturing at which it excelled. It was a world leader in steam engines, sawmills, and flour milling machinery. And, depending on the buyer's wishes, Edward P. Allis and Company could supply one item or a turnkey package designed, manufactured, shipped, and set up at the customer's location.

Bigger things were soon to come. William J. Chalmers, president of the Chicago firm, Fraser and Chalmers, happened to meet Edwin Reynolds of the Allis company in a Chicago hotel. This chance meeting eventually resulted in the merger of several companies to form the historical Allis-Chalmers Company. Articles of incorporation were filed May 8, 1901.

The new company's combined line of heavy machinery for domestic and worldwide trade was most impressive and extensive. About this time it also entered the then-embryonic field of steam and hydraulic turbines. The only big item that it didn't have the facilities for or expertise in building was electrical generating equipment.

The D17 was the first tractor the company produced that came standard with a three-point hitch. Power steering and a dry-type air cleaner were standard equipment, too. This tractor is a 1966 D17 Series IV. Up until the Series IV with 52.7-horsepower, if the purchaser wanted a three-point hitch he had to buy a conversion kit. Having this option available meant he could use other brands of equipment on his tractor if he wished.

73

This D12 is a 1960 D12. Its axles can be adjusted out far enough that it can work two rows. Tricycle and single-front wheel tractors were no longer in demand, so the D12 was offered with only a wide, adjustable front axle.

The company brought this into its fold by first leasing, and then buying outright, the Bullock Electric and Manufacturing Company of Norwood, Ohio. The Allis-Chalmers trademark or logo then appeared claiming, "Ours the Four Powers: Steam, Gas, Water, Electricity."

A promising beginning for Allis-Chalmers led to 10 years of less-than-impressive profits. When the new year rolled around in 1912, the company was bankrupt. Delmar W. Call and Otto H. Falk were appointed receivers of the firm on April 8, 1912. The company was reorganized and incorporated as the Allis-Chalmers Manufacturing Company on April 16, 1913. Falk was its new president, and his brother, Herman, chaired the board of directors.

Tractors were all the rage with manufacturing firms across America in 1913 when Allis-Chalmers Manufacturing Company arose from the ashes of the Allis-Chalmers Company. Fortunately, General Falk was a "tractor man."

Promoting "New for the '60s" tractors, this brochure zeroes in on the new "traction boosters" and how this advancement will help the farmer cover more acres both faster and more efficiently.

As a 1961 model, this D15 didn't have a series designation. It has 40 horsepower and live PTO, but not live hydraulics. As a result, pushing down the foot clutch would also disengage the hydraulics, but the hydraulics still functioned if the hand clutch was engaged instead. The hand clutch provided both high and low ranges, in essence transforming the four-speed transmission into an eight-speed transmission.

Here's where the D15 really shined: operating fertilizer spreaders and similar lighter equipment at ground-eating speeds, all the while providing a comfortable operating platform for the driver.

The company entered the agricultural tractor market in November 1914 with the Model 10-18. This was a three-wheeler with two 56-inch rear-drive wheels and a single front wheel, offset to the right of the vehicle. This allowed both the rear-drive wheel and the offset front wheel to run in the plow furrow. As the model number indicated, the tractor made 10 horsepower at the drawbar and 18 at the belt. The engine was a horizontally opposed two-cylinder that started on gasoline and then switched to kerosene after the engine warmed up.

In 1918, the Model 10-18 was joined by the 6-12, a front-wheel drive design also known as the General Purpose Model A. This tractor was designed for the small farmer who wanted to use the existing horse-drawn equipment he already owned. Model A production ceased in 1926.

In 1918, the company introduced a four-wheeled model, the 15-30. This was changed to the E18-30, based on new test results concerning power. Periodic horsepower increases led to further name changes, such as E20-35 in 1923 and E25-40 in 1929. The latter model was painted Persian Orange.

In 1920, the 12-20 and 15-25 were introduced for a run through 1927. They were also known as the Model L. Original power was from a Midwest motor classed as 15 horsepower at the drawbar.

Allis-Chalmers began building the Model U for the Chicago firm of United Tractors and Farm Equipment in 1929. This firm sold the tractors through its co-op's outlets as the "United" tractor, but the co-op folded and Allis-Chalmers took over the line.

The row-crop Model U served as the basis for Allis-Chalmers' response to the Farmall. It went under the names of All-Crop, Model UC, or Cultivator, and was essentially a Model U with a tricycle front end. It came into the line in 1930.

Following the success of the rubber-tired Model U, Allis-Chalmers engineers developed the Model W All-Crop. This became the Model WC by the time production began in 1934. The WC's high horsepower-to-weight ratio elevated Allis-Chalmers to among the best-known names in agricultural tractors.

"GENERAL FALK"

Otto Falk had his own company that manufactured large machinery, the Falk Company of Milwaukee, Wisconsin, where he'd gained experience in heavy manufacturing. Falk retired from service in the Wisconsin National Guard in 1911 with the rank of brigadier general. He then headed the Allis-Chalmers Company until 1932.

Under General Falk's command, several satellite plants were closed and their production consolidated at the West Allis complex. By 1923, there were only two of the original six plants remaining.

Once Falk got rid of the deadwood facilities, production from the remaining two plants outstripped what had been formerly produced by all six plants. Allis-Chalmers was positioned to once again expand by acquisition.

From 1924 until Falk's retirement in 1932, 10 additional companies became part of the Allis-Chalmers family:

• Worthington Pump and Machinery Corporation added a line of mining, crushing, cement, and creosoting machinery.

• A line of shovels aimed at the mining industry was part of the Hoar Shovel Company purchase in Duluth, Minnesota.

• Flour milling machinery was the main business of Nordyke and Marmon Company of Indianapolis, Indiana, which Allis-Chalmers purchased in 1926.

• Pittsburgh Transformer Company, Pittsburgh, Pennsylvania, became part of Allis-Chalmers in 1927.

• Monarch Tractor Company of Springfield, Illinois, was purchased in 1928.

• LaCrosse Plow Company of LaCrosse, Wisconsin, came on board in 1929.

• Stearns Motor Company joined the company holdings in 1930.

• Advance-Rumely Thresher Company, located in LaPorte, Indiana, was bought in 1931.

• Birdsell Manufacturing Company of South Bend, Indiana, was another 1931 acquisition.

• Hy-Way Machine and Manufacturing Company of Omaha, Nebraska, joined the company in 1931 too.

The only difference between a D12 and this 1961 D10 is that the D12 front and rear axles can be adjusted wide enough to cultivate two rows at a time. This enabled the nonadjustable D10 to carry a lower price tag than the D12 for the economy-minded. Horsepower is 28.4. This increased to 33 with the later Series II models. It's carrying a 60 Series two-bottom plow.

The tractor on the left is a 1960 D12. It has its axles adjusted in as tightly as possible, and it has only 28 horsepower. The unit on the right is a 1963 Series II D12 tractor with axles adjusted out to their full width and powered by a 33-horsepower engine.

77

This is a 1963 D12 Series II tractor. It provides 33 horsepower and the flexibility of axle adjustments from extremely narrow to wide enough that two rows can be handled. Series II designation was added in early 1963 at serial number D12-5501, but no mechanical changes accompanied the new designation.

Beginning with the Model B in 1937, Allis-Chalmers tractors were styled by industrial designer, Brooks Stevens. The RC joined the Allis-Chalmers' line in 1938.

For those who didn't want or need a row-crop tractor, the WF was introduced in 1937 as a standard version of the WC. It was marketed until 1951.

The C, introduced in 1940, was one of five new tractors for Allis-Chalmers during the 1940s. The other new offerings included the IB in 1946, the WD and G in 1948, and the CA in 1949. The WD's foot brakes, hydraulics, and live PTO were important new improvements. The Model G was probably the only successful rear-engined tractor of any manufacturer.

Perhaps the biggest feature introduced in this period was Power-Shift rear wheels, which allowed tread width to be adjusted without the usual jack,

The front axle of a D10 is relatively simple in comparison to that of a D12. That's because the D10 was designed to have nonadjustable axles and be produced and sold as cheaply as possible. Otherwise, a D10 is a twin to the D12.

hammer, and profanity. Debuted on the WD, this Allis-Chalmers invention later became common on most tractors, regardless of manufacturer.

The WD45 came out in 1953 with 25 percent more horsepower than its predecessor, the WD. This put it in the four-plow category. A diesel version of the Model WD45 was introduced in 1954. The D14, D17, and D17D were introduced in 1957. Several new features upgraded these tractors throughout their lifetime.

The Roll-Shift front axle allowed front-wheel tread-width adjustments to be made without jacking up the front of the tractor. The tractor itself supplied the power. Integral power steering and new seat suspension with location ahead of the rear axle made operating the tractor easier and more comfortable.

Used at the D14's introduction was a new four-cylinder engine of Allis-Chalmers design that put it at the 30-horsepower level at the drawbar. The D17 originally fit in the 46-drawbar horsepower class and was powered by a six-cylinder Allis-Chalmers engine in the diesel version. The gasoline and LP gas models of the D17 used four-cylinder engines.

A major innovation on the D14 and D17 was the new Power-Director hand clutch with high and low ranges. This was supplied in addition to the normal

In contrast, the axles of a D12 were made to allow adjusting widths to as wide as 48 inches in order to be able to cover two rows. Note that steering remains easy as the wheel tread becomes wider, thanks to the double pitman.

The 1963 71.5-horsepower D19 was available with a gasoline engine, as with this tractor, and also in diesel and propane versions. The diesel was the first American-built tractor to come with a turbocharger, and the dry air cleaner was the first offered on any tractor. Not until 1965 was a three-point hitch offered, however, and then only in limited numbers. The four-speed transmission was operated via a high- and low-range clutch, so producing eight speeds. A three-spool valve was at the rear for lift and remote cylinders. Power steering was standard.

79

These are the hydraulic controls and a two-spool valve on the D19, something unique for the industry at the time. This provided more flexibility and control, and enabled the utilization of a plow with two hydraulic cylinders plus the lift arms.

THE 1960S AND ALLIS-CHALMERS

What was the big issue concerning tractors at Allis-Chalmers during the 1960s?

Norm Swinford spent 27 1/2 years with Allis-Chalmers and 2 1/2 years with Deutz-Allis. He retired as project manager for tractors. He saw it like this: "I'd have to think the big issue of the 1960s was the introduction of the John Deere New Generation tractors.

"This impacted the entire industry at the time. I believe that's what turned the industry in the direction it went. John Deere had some problems with those tractors in the beginning, but they were definitely state-of-the-art.

"The very fact that they had a whole new powertrain with planetary final drives, wet disc brakes, differential locks, and other new features influenced the industry from then on," he added. "Our people started working on improved powertrains about that time in the early 1960s, along with the other items. Primarily due to economic considerations, we weren't able to put them into Allis-Chalmers tractors until 1973."

Thirteen years is a long time to play "catch-up" when you realize that Deere and Company brought the New Generation from inception to full production in just seven years.

Why did it take Allis-Chalmers so long? Wasn't management cooperating with the tractor engineering and research?

"That was part of the problem," Swinford said. "It depended on who was at the top of the corporation at the time. Bob Stevenson and Willis Scholl were a lot more sympathetic to our needs, although they weren't all that terribly generous all the time. They were old farm equipment people."

As far as tractors and farm implements are concerned, it seems Allis-Chalmers' corporate structure tended more toward centralization than some of its counterparts. All company engineering divisions reported to one vice president, and all manufacturing divisions reported to another vice president.

foot clutch. The Power-Director was an evolutionary step up from the hand clutch on the WD45, which just stopped the forward travel of the tractor, but allowed the PTO and hydraulic system to stay engaged.

The Power-Director did that, plus added a low range to the transmission. It effectively provided on-the-go shifting to get through the tough spots more readily.

RACE TRACTORS

Race tractors? Yes, an old advertising gimmick with a new twist. Early automobile manufacturers or dealers, including Henry Ford and Harry Ferguson, put race cars on the circuit to promote their products. Allis-Chalmers borrowed a page from their book in the early 1930s.

In 1932, the Allis-Chalmers Model U had the distinction of being the first farm tractor offered with pneumatic rubber tires. Testing that proved the feasibility of rubber air tires began with smooth-tread airplane tires running only 15 pounds per square inch.

Next Allis-Chalmers consulted with Firestone to design an air tire especially for farm tractors. The rest is history.

It took a lot of promotion and persuasion to convince farmers that the extra $210 initial cost was worth the money. Some of this promotion and persuasion came in the form of tractor races held at local fairs across the country.

A special fourth gear was installed in the transmission of several "race circuit" Model Us. In Dallas, Texas, in 1933, famed race car driver Barney Oldfield piloted a Model U at 64.28 miles per hour.

The top tractor speed record was later set by Ab Jenkins on the Utah salt flats at 67.877 miles per hour also driving an Allis-Chalmers Model U.

It isn't certain whether it was the race publicity or actually using rubber tires in the field that convinced farmers. Regardless, rubber tires on farm tractors were here to stay.

Due to the large scope of the company and the level of diversification, it was "badly unstructured" according to Swinford. Finally, in the early 1960s, the farm equipment division was organized under its own vice president.

Two other areas where Allis-Chalmers differed from many of its counterparts stand out. First is that there was never an attempt to standardize tractor design worldwide. Second, although Allis-Chalmers had facilities in many parts of the country, tractors were never produced at any U.S. facility other than West Allis. The sole exception was the Model G, built in Gadsden, Alabama.

Metal casting had always been a strong suit of Allis-Chalmers dating back to 1847 when Allis purchased the Decker and Seville facility. As the company grew through the years it acquired the expertise, equipment, and facilities to make enormous castings that were among the largest in the world.

A turbine runner for the Niagara Falls Power Company was cast by Allis-Chalmers. This runner measured 15 1/2 feet in diameter, weighed 50 tons, and was cast in a single piece.

The West Allis plant had its own foundries until environmental issues caused Allis-Chalmers to rethink the economics of maintaining its own

The D21 was a milestone tractor, because it was the company's first tractor to offer more than 100 horsepower. This 1964 D21 had 103 horsepower. Using both inner and outer weights, dual wheels, and fluid in the tires, the tractor's weight could be brought up to 17,000 pounds for better traction. It was more popular in the Western states than in the Corn Belt.

The dual-equipped D21 had enough power that it might have to be dropped into a slower gear to keep from overtaking the tractor doing the disking or the combining, even though the chisel's shovels are socked well down into the soil.

foundries. One of the two foundries was closed before 1960 to become part of the tractor plant in 1963. The second was closed some time later. Allis-Chalmers then began securing large casting from outside firms, a practice known as outsourcing.

The West Allis facility was world headquarters for all Allis-Chalmers activities, including offices for the international division. It was also engineering's home base.

How big of a part did tractors really play at the West Allis facility? Swinford noted that, "I can't really say that tractors were a big part of West Allis. Only approximately 25 percent of the floor space was devoted to tractors."

West Allis had extensive test facilities at the complex, as well as a proving ground in Racine County. Like other manufacturers' tractors, Allis-Chalmers prototype tractors were "snowbirds." That is, they were shipped south to Texas for winter field testing.

How did all this affect Allis-Chalmers tractors in the 1960s?

Several of the D series tractors that were in production in the 1950s made the transition to the 1960s. Along the way, these D models received new features and new styling before they were eventually phased out of the line.

In 1960, two new models of the D appeared, the D15 and the D15D. A four-cylinder gasoline, diesel, or LP gas engine powered these tractors to the 35-drawbar, and 40-PTO, horsepower level.

These tractors were followed the next year with the D19 Model. The D19 diesel took a giant step forward for the company when it was introduced equipped with a turbocharger that put it in the 65-horsepower class.

The turbocharger was standard with the Allis-Chalmers six-cylinder engine available in gasoline, diesel, or LP gas. The D19 transmission provided eight forward gears. Another noteworthy feature was the dry-type air cleaners that used precleaners and automatic dust unloaders.

The next year, 1963, the D21 debuted with new styling that made the rest of the D series units appear rather dated thanks to their styling, which had been around for seven years. A full bevy of new features accompanied the D21, but the biggest thing that it gave the Allis-Chalmers line was its first tractor to break the 100-horsepower mark.

Its new engine was the Model 3400 diesel featuring direct-injection—an Allis-Chalmers first. This gave the D21 horsepower ratings of 103 PTO and 93 on the drawbar.

The powertrain was new and much larger and heavier than that of the D19. Thanks to the Power-Director, the transmission allowed eight forward and two reverse choices.

This was also Allis-Chalmers' first tractor to have hydrostatic power steering. It came with two 12-volt batteries for surer starting even in cold weather. The powerplant was Allis-Chalmers' six-cylinder 426-cubic inch engine. The fuel tank was enlarged to 52 gallons for longer field time between refills.

Not wanting to be left behind in the four-wheel-drive race, Allis-Chalmers introduced its "Sugar Babe" in 1963. Actually, the Sugar Babe was the Model T16 that acquired its unusual name from use

In 1963 and 1964, the D21 tractors had chrome grilles like that of this 1964 tractor. The following Series II D21s had white or cream-colored painted grilles. These later tractors also came with turbochargers and intercoolers that boosted horsepower up to 124.

in south Florida's cane country. How it ended up toting sugarcane isn't an impressive story for Allis-Chalmers' third four-wheel-drive effort.

Its second effort at building a four-wheel-drive vehicle occurred circa 1945 with the Model H. Six prototypes were tested extensively, but they all failed due to a fault in the steering system. It was a good idea, but ahead of the period's engineering technology, and so it never saw production.

You have to go way back to the Model 6-12 to find Allis-Chalmers' first four-wheel-drive tractor. This unit was developed between 1918 and 1926. Its four-wheel-drive design consisted of two Model 6-12's hitched in tandem. It was called a Duplex and was capable of handling a three-bottom plow. Not many were manufactured and few survived.

The Sugar Babe is about as rare since only 15 vehicles were sold by the farm equipment division. Some of these ended up in the wheat-producing country, although they didn't work out and were bought back by the company. The T16 was built around the TL16 loader from the construction machinery division's Deerfield, Illinois, plant. It suffered from two major problems when used in agricultural application. The original four-cylinder turbocharged diesel engine wasn't tough enough for hour-after-hour fieldwork under continuous load. When the four-cylinder was swapped for a six-cylinder engine, torque converter problems began to surface; however, they proved well fitted to hauling sugarcane, hence the name "Sugar Babe."

Allis-Chalmers' first all-new line for the decade came with the 100 Series. The first was the One-Ninety launched in 1964, followed by the One-Ninety XT the following year.

The XT model was similar to the One-Ninety, but squeezed 16 additional horsepower from the engine thanks to a turbocharger. Both the One-Ninety and One-Ninety XT came in the buyer's choice of gasoline, diesel, or LP gas models.

"Built Like a Bull!" seems much more apropos in describing the tractor than the polyethylene tanks. The One-Ninety XT model had the brawn to handle the fieldwork on most farms when it was introduced in 1964 for a 10-year production run. This particular tractor was an early one in using dual wheels to lessen yield-damaging soil compaction.

The year 1967 added the One-Seventy and the One-Eighty. Power steering was upgraded from the power-assist type on the D17 to a hydrostatic system. The gasoline model continued with some of the D17's features such as Power-Director shift-on-the-go eight-speed transmission and continuous 540 PTO. The Snap-Coupler hitch system, introduced on the WD45 in 1953, was discontinued in favor of a category II three-point hitch. The diesel model featured a four-cylinder Perkins 236-cubic inch engine. Due to its waning popularity, LP gas wasn't available on either tractor.

CONTINUED EXPANSION

After the flurry of company acquisitions under General Falk at Allis-Chalmers, another round of expansion got under way during the late 1940s and continued into the 1980s. Below are some of the acquisitions of the 1960s, with date of purchase and some products produced:

Year	Company	Location	Products
1948	Gadsden plant	Gadsden, AL	Cotton pickers
1950	Essendine plant	Essendine, England	Harvesters, tractors
1951	Canadian Allis-Chalmers	Lachine, Quebec, Canada	Heavy equipment
1952	LaPlant-Choate Mfg. Co.	Cedar Rapids, IA	Motor scrapers
1953	Buda Company	Harvey, IL	Engines, lift trucks
1955	Gleaner Harvester Corp.	Independence, MO	Combines
1955	Baker Company	Springfield, IL	Bulldozers
1957	T.C. Pollard Pty. Ltd.	Newcastle, Australia	Motor graders
1958	Industrial Dufermex, S.A.	Mexico City, Mexico	Transformers
1958	Micromatic Hone	Detroit, MI	Diesel engines
1959	S. Morgan Smith	York, PA	Hydraulic turbines
1959	Tractomotive Corp.	Deerfield, IL	Wheel loaders
1959	Allis-Chalmers Italiana	Cusano, Italy	Crawler tractors
1959	Valley Iron Works Corp.	Appleton, WI	Paper machinery
1960	Establissements de Constructions de Vendeuvre SA	Dieppe, France	Engines, generators
1963	Schwager-Wood Company	Portland, OR	Electrical equipment
1963	New factory	Guelph, Ontario, Canada	Loaders, lift trucks
1963	New factory	San Luis Potosi, Mexico	Lift trucks, combines
1965	Simplicity Mfg. Co.	Port Washington, WI	Lawn and garden
1968	Henry Mfg. Co.	Topeka, KS	Industrial tractors
1969	Standard Steel Corp.	Decatur, IL	Asphalt plants
1969	New factory	Little Rock, AR	Electric motors
1969	New factory	Matteson, IL	Material handling equipment

This attention-getting brochure almost makes the reader feel the big, turbocharged D-21 is ready to roll onto his lap. The headline message, "Year 'Round Work for the Man Who Farms BIG," leaves no doubt that this is a hard-charging tractor for an equally hard-charging farmer with lots to do.

Allis-Chalmers' One-Sixty Diesel was offered in 1969. It fell in the 40-horsepower range and offered 10 forward gears. It was built to Allis-Chalmers specifications by Renault in France and featured a three-cylinder Perkins engine. It replaced the D15.

The year 1969 brought the Two-Twenty into the line with eight more horses than the D21 Series II that it replaced. It was rated at 136 horsepower and was powered by Allis-Chalmers' 426-cubic inch six-cylinder turbocharged engine. Transmission was four-speed with a hi- o range that provided eight forward and two reverse speeds.

Under the category of imported tractors, the ED40 built in Essendine, England, was brought into Canada. Eventually, 450 crossed the border into the United States. The ED40 was a product of the international division, which had its own manufacturing and engineering.

Pulling a 12-row corn planter with ease, the One-Ninety XT had power to spare. Note use of dual wheels to limit compaction and one of the earlier protective tractor cab models. The One-Ninety gas and diesel versions only rated 75- and 77-PTO horsepower, respectively—a bit underpowered for both the sales staff and the farmer. The engine division responded with a larger 301-cubic inch turbocharged diesel, gasoline, and LP engine. Fitted with the new engine the One-Ninety became the One-Ninety XT that jumped the PTO horsepower to 93.

What sets this 1965 Series III D12 Hi-Crop apart from other D12s are its taller spindles and different, bigger rear wheels and tires. All D10s and D12s of this mid-1960s vintage put out the same 33 horsepower.

Was the beginning of the end for Allis-Chalmers rooted in management's perception of the farm industry as early as the 1960s? Perhaps, as indicated by this reflection from Swinford: "David Scott became president in 1968. He was an elitist and farm equipment was beneath him except during the 1970s when it made a lot of money, because everybody could sell tractors.

"Then, when things went sour, he went sour on farm equipment. When it became time to start spinning off elements of the company, he was quick to start thinking about farm equipment."

THE CYCLE CONTINUES

The 100 Series gave way to the 7000 Series beginning in 1973. Some models stayed in the line until 1981, while others faded out somewhere in between. The 6000 and 8000 tractors replaced the 7000 line and were the last of the pure Allis-Chalmers line.

In 1980, Allis-Chalmers' worldwide scope included 71 plants, plus 100 branches, sales offices, and

DEFINING DIVERSIFIED

Allis-Chalmers tractors benefited from the unparalleled amount of manufacturing expertise the company acquired in many diverse fields long before there were any tractors in the firm's catalogues. The demise of Allis-Chalmers tractors and farm equipment may well have been the result of this same diversification.

The Allis-Chalmers Company could serve as an accurate dictionary definition of "conglomerate." Throughout its history it was involved in, and excelled at, scores of industrial manufacturing enterprises.

Below is a list of some of the product areas the company explored to show the company's scope. As you read the great number of product areas, also think of great size—big, actually gigantic, products were its forte as shown by a few examples:

In 1919, Allis-Chalmers manufactured and installed the world's largest, at the time, combined hydroelectric project at Niagara Falls. Initially the power of the hydraulic turbines was given as 70,000 horsepower, but it was later rated at 75,000 horsepower.

Allis-Chalmers built some of the world's largest kilns. Some required either special rail cars or as many as four or five rail cars to accommodate their enormous size in transit to customers' facilities.

In 1966, the company fabricated a steel chain with 1x3-inch links, 29,000 feet long and weighing 375,000 pounds. It was used in a cement kiln.

Allis-Chalmers constructed large 12-cylinder marine engines capable of 43,200 horsepower in the 1980s. The bedplate of each was 80 feet long and weighed 80 tons. Each cylinder jacket weighed 15 tons. Total weight of the engine was 1,160 tons.

Other items the company produced included
- Blowers, compressors, and vacuum pumps
- Cement machinery
- Condensers
- Steam engines
- Crushing machinery
- Electrical machinery engines
- Farm machinery
- Forgings
- Flour mill machinery
- Hoisting machinery
- Metallurgical machinery
- Plate work
- Power transmission machinery
- Pumps
- Sawmill machinery
- Timber preserving machinery
- Tractors
- Steam turbines
- Hydraulic turbines
- Water softeners
- Home heating systems
- Refrigerators
- Paper-processing equipment
- Outdoor and recreational equipment
- Lawn and garden tractors
- Ore and mineral processing equipment
- Nuclear power
- Industrial and construction machinery
- Military production
- Kilns
- Mining machinery
- Fuel cells

The rear end of the 1965 Series III D12 Hi-Crop provides both a three-point hitch and a small drawbar hitching arrangement.

Built on a D10-/D12 chassis for industrial usage, the I40 used a heavier front end and heavier spindles, although the front end was nonadjustable. It had a 3/8-inch plate grille, instead of a radiator shell, that was set forward. This allowed the hydraulic pump to fit between the radiator and the grille for live power and more hydraulic capacity than a comparable farm-type tractor. This feature let the I40 mount backhoes, blades, and front-end loaders. This is a 1964 model.

warehouses staffed by 29,000 employees. Approximately 5,000 dealers sold and serviced its products, which generated almost $2 billion in sales during 1979.

Whatever the theory for the demise of Allis-Chalmers, the end result was that on March 28, 1985, Allis-Chalmers was sold to Klockner-Humboldt-Deutz AG of Germany.

The transformation of Allis-Chalmers continues to evolve under the banner of AGCO. Fiat-Allis is the badge on construction equipment. The electrical equipment division was folded into Siemens-Allis.

Products of the agricultural equipment division became Deutz-Allis in 1985. Then, in 1990, AGCO Allis became part of AGCO (Allis-Gleaner Corporation) along with White Tractor, Hesston, GLEANER, Deutz-Allis, SAME, and Massey-Ferguson. Today AGCO manufactures and distributes the world's largest line of agricultural machinery.

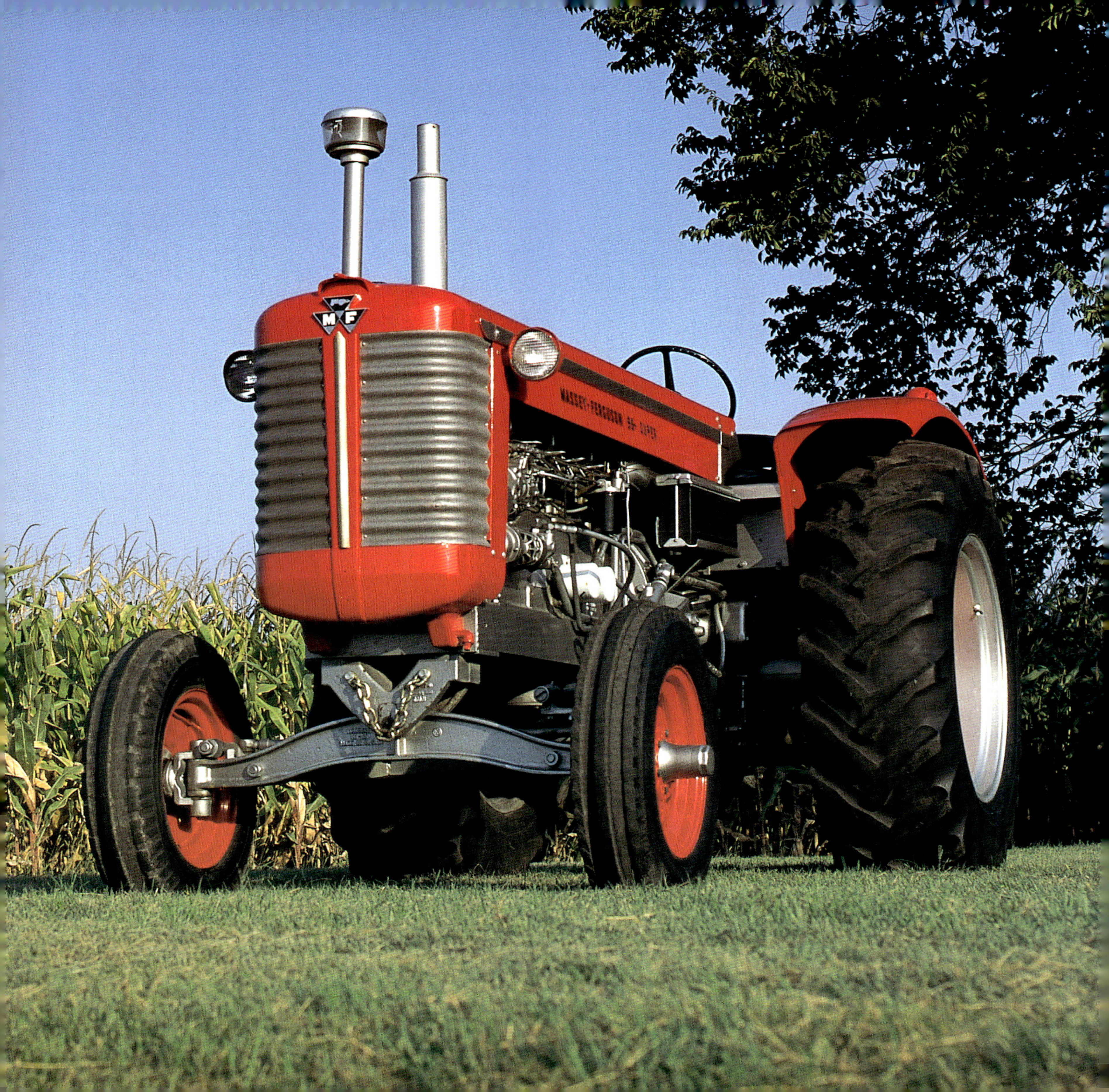
MASSEY-FERGUSON

CHAPTER FIVE

MASSEY-FERGUSON, INC.

The roots of the Massey side of the Massey-Harris-Ferguson family tree can be traced back to 1847, when Daniel Massey began to manufacture farm implements at a plant in Newcastle Province of Canada West (now Ontario). Massey incorporated as Massey Manufacturing Company in 1870 and moved his growing business to Toronto nine years later.

Alanson Harris entered the farm implement manufacturing arena 10 years later, in 1857, in Beamsville Province of Canada West. In 1872, he relocated his business to Brantford, Ontario, Canada.

Massey and Harris were the leading suppliers of reapers, mowers, and harvesting equipment north of the border. After years of competing for the Canadian market, the two merged companies in 1891 to form the Massey-Harris Company Ltd. in Toronto. Almost immediately the new company started purchasing many of its rival companies. The Patterson Wisner Company, Verity Plow Company, Corbin Disc Harrow, the Bain Company, Kemp Manure Spreader Company, Deyo-Macey Company, Johnston Harvester Company, and J. I. Case Plow Works were all eventually brought into the Massey-Harris holding. Purchase of the Case company, in 1928, was the last significant purchase or merger for the following 25 years.

Massey-Harris' first venture into the tractor market was the Big Bull in 1917. The company imported it from the United States, where it was sold by the Bull Tractor Company of Minneapolis, Minnesota.

This was followed in 1919 by the first Massey-Harris–built tractor, patterned after the product of the Parrett Tractor Company of Chicago, Illinois.

This tractor, manufactured at the Massey-Harris factory in Weston, Ontario, was identified as the Massey-Harris #1, #2, and #3.

Like the Big Bull, this venture was a failure. In 1923, Massey-Harris moved temporarily out of the tractor business, but to be competitive as a full-line farm equipment company, tractors were a must. So in 1927, Massey-Harris obtained the rights to market the J. I. Case Plow Works–built Wallis Certified three-plow tractor in Canada and some parts of the United States. A year later, Massey-Harris purchased the entire J. I. Case Plow Works Company, giving it total control of the Wallis line.

In 1930, the company offered the Massey-Harris General Purpose four-wheel-drive tractor. This innovative vehicle was perhaps three decades ahead of its time, though. It came with adjustable tread widths, but was a backdoor approach to the row-crop tractor that didn't work.

Massey-Harris brought out the 25-40, also called the Model 25, in 1931. This was a three-four plow tractor that didn't do well in either the marketplace or the field.

The company's next offering was the Massey-Harris Pacemaker. It was introduced in 1936 as a three-plow, gasoline standard tractor available on the buyer's choice of steel wheels or rubber tires. Horsepower was rated 16 at the drawbar and 27 at the belt.

Also in 1936, the Massey-Harris Challenger joined the family as a row-crop or standard unit. The Challenger could also be delivered with steel or rubber adjustable back wheel tread. The four-cylinder engine was an all-fuel type. The Challenger was at the 26-drawbar and 36-belt horsepower level.

Powered by a 74-horsepower engine, the MF 95 Super was designed to pull a five-bottom plow. The Super series, such as this 1961 model, was built from 1960 through 1963. Interesting note: The 95 Super was actually made by Minneapolis-Moline and is almost the same as that company's Model GVI.

89

The bulge in this 1960 88 LP tractor's topline is the LP fuel tank that was needed to provide more capacity than if a diesel or gasoline was installed instead. The engine developed 60 horsepower at the drawbar, and there was no three-point hitch. Its owner says that "This was the best-handling tractor of the era, and it was easy to get on and off, because you could do that from the rear or either side."

Ordered from Oliver in 1959 for a run of 500 tractors, and the same tractor as the Oliver 990, this 1960 98 diesel was one of the just 400 actually manufactured. Power is from a 371-cubic inch diesel engine and is 80 horsepower.

The Massey-Harris Model 101 was a two-three plow tractor introduced in 1939. It featured a Chrysler-built six-cylinder engine based on Chrysler's automobile engine design. The 101 was available on rubber or steel wheels that could be adjusted from 52 inches to 90 inches on some models.

Across the Pacific, Harry Ferguson built the prototype of his first tractor in 1936. Soon thereafter he contracted with David Brown of Huddersfield in Yorkshire, England, to manufacture tractors using his revolutionary hydraulic three-point hitch system. The resulting tractors carried the Ferguson-Brown name.

These tractors proved the value of Ferguson's hydraulic system and three-point hitch with draft control; however, it wasn't until he teamed up with Henry Ford in the United States that his idea was transformed into a successful commercial tractor using the Ferguson system. More than a half million Ford-Ferguson tractors were sold.

Still, Ferguson didn't have his own tractor—not that he really ever wanted to be in the manufacturing business; but he did want to be in control. And as long as Ford was involved he wasn't in control.

In 1946, Ferguson and Sir John Black of England's Standard Motor Company reached an agreement that allowed Standard Motor to manufacture

90

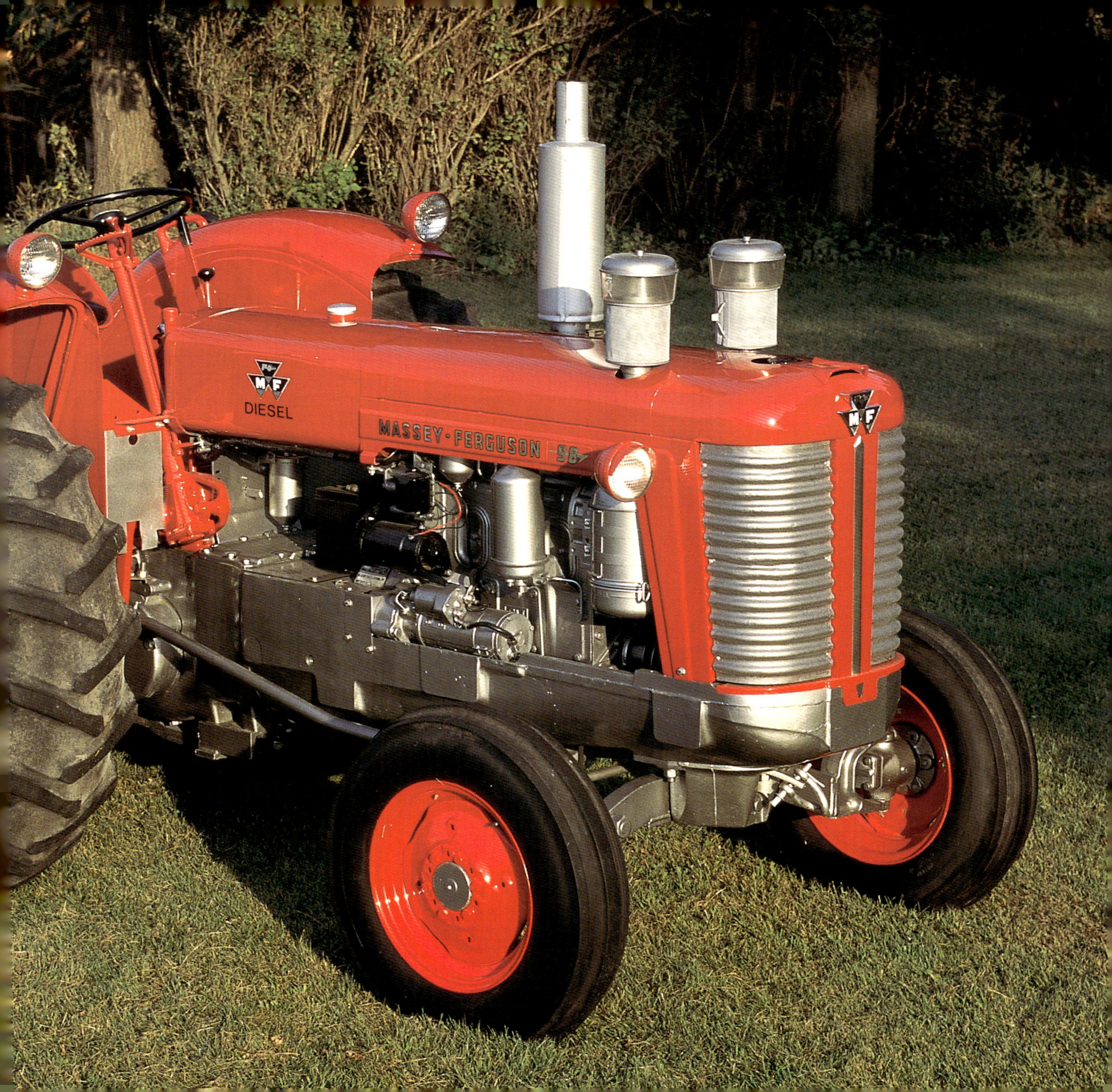
M F
DIESEL
MASSEY-FERGUSON 98
M F

It's doubtful that the factory workers finishing assembly of new Massey-Ferguson tractors actually wore such sparkling white—and clean—outfits all the time. Regardless, it was a fitting last page for a 1960 calendar given out by dealers to customers, past and potential.

Ferguson's TE-20 tractor. Also in 1946, Ford ended the "handshake" agreement to manufacture tractors for Ferguson. The last Ford-built tractors were delivered to Ferguson in July 1947. Once again Ferguson found himself without a manufacturer for his tractor in the United States.

If Massey-Harris tractor production wasn't brilliant in the 1940s, the company did have a rising star that would finally give the company strong name recognition in the United States. The "Massey-Harris harvest brigade" was conceived in 1944 and proposed to put an additional—above war quotes—500 new Massey-Harris self-propelled combines in the hands of custom combine operators.

Emerging from World War II, the company had less than half the worldwide market share of Deere & Company, but the market was insatiable. A further boon to Massey-Harris was that it was the only major manufacturer not to suffer a labor strike in the postwar boom years.

In 1949, Massey-Harris made the Hydraulic Depth-O-Matic Control available on the Massey-Harris 22 as special equipment. This same hydraulic feature was available as special equipment on the Massey-Harris 30 Pony and 44 for the 1950 model year. This was actually a poor and tardy response to Ferguson's innovative system. It didn't include three-point linkage or weight transference, and the hydraulic system was an external attachment instead of built into the tractor.

A year later, in 1951, a three-point hitch was available on the #22 Row-Crop models. Yet it still wasn't anything close to the Ferguson system that was getting the lion's share of the market in the two-plow, small tractor market. Massey-Harris felt this justified investing a great deal of time and money to develop a tractor line that could go head-to-head with the Ferguson system; however, budget cuts in 1952, and a prototype flop known as the Pitt tractor in 1953, kept the tractor program in serious arrears of the competition.

By 1951, Massey-Harris controlled 51 percent of the self-propelled combine market in the United States. It was a much different story with tractors and implements, though. In that same year, tractors only registered 4.8 percent and implements 3.0 percent of the U.S. market.

Several far-reaching decisions concerning tractors occurred at Massey-Harris during the decade of the 1950s. None, however, was more important than when Massey-Harris bought Harry Ferguson's company in 1953.

Ferguson was given an honorary position on the board of directors; however, he quickly demonstrated the personality traits that caused Henry Ford II to comment that "Ferguson was a difficult man to do business with." These "difficulties" ended the relationship between Ferguson and Massey-Harris only a year later, in 1954. An agreement terminating Harry Ferguson's association with Massey-Harris-Ferguson Ltd. was dated July 6, 1954.

After the "merger" the Massey-Harris tractor plant at Racine, Wisconsin, closed and all tractor manufacturing was consolidated at the Detroit factory where the Ferguson tractors were produced. Some of the property at the Racine location was sold. The remaining facilities became a central parts warehouse known as the North American Parts Operation.

Perhaps most significant, in respect to tractor production, was the decision to make as many components as possible interchangeable between tractors produced in any of Massey-Ferguson's plants. Though it was never fully achieved, the goal helped to reduce costs and duplication of engineering and manufacturing efforts.

In 1959, Massey-Ferguson purchased F. Perkins Ltd. of Peterborough, England. This gave the company its own engine production capability, which it had lacked since the old J. I. Case Plow Works in Racine stopped manufacturing engines in the late 1930s. In the intervening years Massey-Harris had outsourced all its tractor engines.

Although basically the same as the 35 farm tractor, this 1961 202 is the industrial version; however, it was built with a heavier frame, heavy nose, had additional weight at the rear end, and used a foot accelerator. Some industrial units also featured a torque converter in the transmission. The engine provided 40 horsepower.

In 1962, this 35 tractor was manufactured. It was a utility model that provided 35 horsepower, or enough for light-to-medium fieldwork and around the farmstead jobs. F. Perkins, Ltd. of Peterborough, England, was bought by Massey-Ferguson in 1959. After the purchase, many of the three-cylinder Perkins diesel engines were used in the Massey-Ferguson Model 35 tractors.

This 1962 Massey-Ferguson 65 left the assembly line in 1962. The diesel engine installed in it produces 51 horsepower. A gas model 65, however, was rated at 3 horsepower less. A four-cylinder Continental was used in the gasoline and LP gas models, while a Perkins four-cylinder powered the diesel tractors.

WORLD'S LEADING ENGINE MANUFACTURER

In 1968, Massey-Ferguson literature gave this overview of the company's engine program:

In line with its policy of manufacturing the greater part of its product, Massey-Ferguson purchased F. Perkins Limited of England in 1959, now the world's largest maker of diesel engines, producing some 300,000 units a year. More than 850 manufacturers around the world specify Perkins diesels as original equipment; it is the second largest supplier of diesel engines for U.S. pleasure craft.

Perkins' largest plant at Peterborough, England, has a capacity of 1,400 units a day. Other plants are in Australia, Brazil, and France. Producing associate companies are in Argentina, Mexico, and Spain, and licensees are in India, Italy, and Japan.

Throughout the 1960s many expansions were initiated to increase and upgrade production. These included

1964—The Peterborough plant was expanded, resulting in a 15 percent increase in sales to $94 million.

1965—A new plant was acquired in the United Kingdom to produce V-8 engines and a new four-cylinder model for automobile application.

1968—An announcement of a new $25 million plant in Turkey was made. Initial capacity was geared for 50,000 engines per year. Expansion of U.S. and Canadian engine facilities was planned.

MASSEY-FERGUSON IN THE 1960S

Why was it important, even crucial, for farm equipment companies to maintain a strong emphasis and presence in foreign markets?

In North America, the number of tractors used for agriculture increased only 8 percent from 1954 to 1964. On the other hand, in developed countries as a whole the number increased 58 percent. In developing countries the number of units in use increased by 106 percent.

During the 1960s the numbers looked good for Massey-Ferguson in its foreign markets outside of North America, but in 1965 Massey-Ferguson tractors only accounted for 13.5 percent of the U.S. tractor market.

The company's tractors followed John Deere, International Harvester, and Ford; however, at 23 percent, Massey-Ferguson did account for the largest share of the Canadian market.

The 51 horsepower of this 65 Diesel could be used for such mundane tasks as towing the manure spreader or pulling the feed wagon, or for some relatively heavy usage. Note this tractor's high clearance.

For the rest of the world, except for Communist countries and India, Massey-Ferguson tractors were number one with approximately 20 percent of the market. Ford, International Harvester, Deere, Fiat, Renault/Porsche, and David Brown followed.

In 1961, the Massey-Ferguson MF-25 was introduced from a French factory. Fitting into the 25-horsepower class, it was offered with a four-cylinder diesel-only engine. Unfortunately, the engines produced at the St. Denis facility (the former Standard-Hotchkiss facility) were troublesome.

Del Gentrer retired from Massey-Ferguson after 43 years of service. His career included working as a service rep, service manager, and district manager. From these positions he went into product training, service training, and warranty service. These are his comments about the MF-25:

"We needed a tractor of that horsepower range—20 to 30 horsepower. It was my understanding that it would have been too expensive to come up with another TO-20 or a TO-30. So the MF-25 was introduced in France and imported into the United States.

"The only minor troubles it had concerned injector and hard starting problems. This was corrected by Thermo-Starts in the manifold. We thought of making sure the customer understood how to start a diesel of that nature. Powerwise, it was close to a Model 35, and in some conditions it would perform equal to the 35."

To deal with the obvious trend toward larger horsepower tractors, especially in the United States, the company decided to plan for an entirely new line of tractors in 1962.

Several features were mandated from the beginning: The tractors would be of large horsepower,

Massey-Ferguson MF 85 DIESEL GASOLINE LP GAS

5-PLOW POWER WITH THE FERGUSON SYSTEM

THE MASSEY-FERGUSON SYSTEM OF MECHANIZED FARMING

Cover of the 85 model brochure showed the tractor easily turning the soil with a five-bottom plow. It was also pointed out that the tractor was available with the buyer's choice of three types of engines: diesel, gasoline, or LP gas.

95

This is quite a rare tractor, because the 1963 97 left the factory with four-wheel drive rather than the conventional rear-wheel drive in most such tractors of the time. The engine develops 100 horsepower.

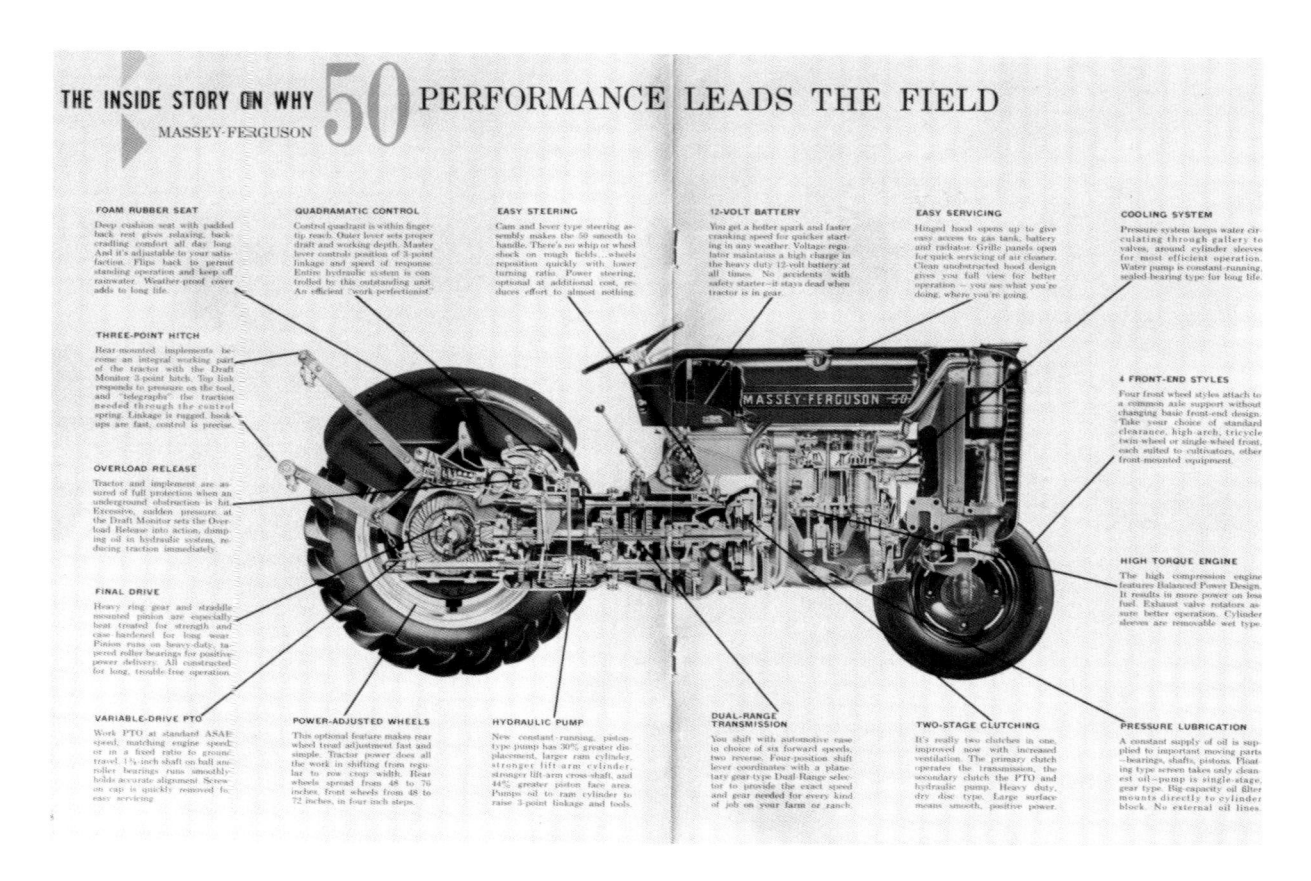

What just well may be the most comprehensive tractor cutaway illustration of the decade of the 1960s describes what sets the Model 50 off from all the competitors in its horsepower and price ranges.

incorporate an advanced version of the Ferguson System, and be designed to use standardized component parts manufactured in different countries.

The new line would include the MF-135, the MF-150, the MF-165, the MF-175, and the large MF-1100 and MF-1130. To move these new tractors from inception to production, the company made a commitment that involved prodigious amounts of time and money.

The engineering division ear-tagged one million man-hours to the design, prototypes, and testing of these new models at a cost of about $7.5 million. Nearly $22 million was invested in capital expenditures for manufacturing. More than $10 million was necessary to develop new engines for the line.

Massey-Ferguson had a big year in 1965 when these new tractors were introduced. All engines were direct injection type as opposed to the previous design. The engines for the MF-135 and MF-165 were improved models of the Perkins, and the MF-175 engine was a completely new Perkins design that featured a built-in balancer to ensure smooth performance.

The MF-1100 joined the industry turbocharging trend with a Perkins six-cylinder turbocharged diesel powerplant that charted 85-drawbar horsepower.

The MF-135, MF-165, and MF-175 all got bigger and better front axles that included larger diameter stub axles, bigger bearings, and improved hub seals.

Priority consideration was given to the many different working conditions these tractors would have to perform in worldwide application. Standard-duty and heavy-duty rear axles were available to meet local requirements.

Transmissions could be the six-speed or the Multi-Power 12-speed. Increased fuel tank capacity was incorporated into the larger models. The fuel tanks of the MF-1100 were a new saddle tank concept that increased safety during fueling and added more fuel capacity.

Operator convenience and comfort were addressed with an improved seat, bigger platform, and additional gauges as standard equipment. Carried over from 1958 was the MF-65 Diesel produced at the Detroit, Michigan, plant. This unit had mid-40s horsepower from a four-cylinder Perkins engine mated to a six-speed transmission.

ENGINEERING AT MASSEY-HARRIS AND MASSEY-FERGUSON

Massey-Harris could lay claim to one major engineering coup during its lifetime: the self-propelled combine. Ferguson could also claim its own engineering coup: the Ferguson hydraulic three-point hitch.

What these two revolutionary engineering accomplishments had in common was the absence of traditional professional engineering methods. In other words, they were the products of intuitive trial and error designed by "hands-on" methods that fit-and-changed and fit-and-changed until the idea worked.

Typically, a lot of the "engineering" was performed in the shop and in the field without reliance on engineering formula, slide rule, and drafting table. The resulting product, or system, from this method is always revolutionary in concept, and if successful can produce great rewards. There is also an extremely high risk of failure.

The other approach to engineering was the bringing of "evolutionary" changes to existing products. This seldom, if ever, resulted in any major advancement; but neither did it carry the high risk of failure. Evolutionary changes were also vital to keeping existing products competitive.

Ferguson engineers were of the former school. Even after the merger with Massey-Harris they continued with this approach, to the neglect of evolutionary engineering changes. Massey-Harris didn't seem to excel at either approach regarding tractors.

Finally, in 1959, the company implemented some much-needed changes in its engineering structure. One obvious need was to align engineering efforts to market research. Somehow this concept, which International Harvester and Deere and Company employed to great benefit, had previously escaped management's attention at Massey-Harris-Ferguson.

In a move to correct this, a Realignment Manual, published in 1959, outlined the new role of engineering at Massey Ferguson. This came at the same time that a new management position was created within the engineering department. A vice-president of engineering was appointed. He was directly responsible for all engineering activities of the company.

The role of the engineering vice-president was to focus engineering efforts in three main areas: First, he was responsible for improving current company products; second, he had the responsibility of designing and developing new and improved models of company products; and third, he had to make a critical analysis of what job the company's products were to perform and how well they were meeting that criterion.

This was an important management decision that helped correct the lack of evolutionary engineering by the Ferguson side and the lack of quality control apparent on the Massey-Harris side of the equation.

Beginning in the 1960 model year, the Diesel-Matic was part of the MF-65 package. The Diesel-Matic designation meant the tractor had the Multi-Power shift-on-the-go transmission and direct-injection engine. The MF-65 was replaced by the Model MF-165 in 1965.

Another late 1950s model carried over into the 1960s was the MF-88 Diesel, introduced in 1959. It was offered through 1961. This tractor was the Wheatland, or Western, version of the MF-85. The MF-88 fell into the mid-50s class for horsepower, which was provided by a four-cylinder Continental engine. It was strictly a pull-type tractor without hydraulics. It came into the line as a replacement for the 333 and 444 series tractors.

Coming on board in 1961, and offered through 1965, the Super 90 Diesel model used a four-cylinder Perkins engine. It was also available with a Continental engine for farmers who preferred gasoline or LP gas. Fitted with an eight-speed transmission, the Super 90 was in the 60-horsepower area. A product of the Detroit plant, it was available as a Wheatland model during 1961 through 1965.

The horsepower race was really spiraling upward by the early 1960s. Some companies were offering 100-horsepower models. Gentner pointed out Massey-Ferguson's position at the time: "In the late 1950s and the early 1960s, we had up to a 63-horsepower tractor which was the MF-85, and in the 60s, the Super 90.

"You see, we needed some bigger tractors, bigger than what we had with the MF-85 and the Super 90. As engineering came up with the new model line it was decided that it was better to purchase bigger tractors from an outside source, Oliver and Minneapolis-Moline, that would carry our sheet metal.

"The Minneapolis-Moline was to be sold in Canada, and the Oliver was to be sold in the United States. But, as it turned out, both tractors were sold in both countries."

This outsourcing decision resulted in the MF-95 Diesel being manufactured by Minneapolis-Moline in 1958. It used the Minneapolis-Moline six-cylinder engine that registered right at 90 horsepower and came with a five-speed transmission.

In 1961, the MF-95 became the MF-97 Diesel. It was the same tractor as the Minneapolis-Moline G 705 Diesel and was available in an LP gas version too. A front-wheel-assist option was available on the MF-97. Of course the sheet metal, paint scheme, and badge were appropriate for the Massey-Ferguson line. The Model MF-97 was offered until 1965.

In 1959 and 1960, approximately 500 tractors were purchased from the Oliver Corporation and sold as the MF-98. This tractor was based on the Oliver 990 chassis with a four-cylinder GM diesel engine that developed 84 horsepower. The sheet metal, decals, and color were appropriate for Massey-Ferguson tractors.

The MF-135 was a close cousin of the Ferguson 20. It replaced the Model 35 in 1964. That year production came from the Detroit plant, and in 1965 it was also produced in the United Kingdom. Production stopped in 1976 at both locations.

Its owner says this 1963 25 will be restored when it's worn out, and not before. A vineyard model, the tractor's narrow tread width is designed to enable it to easily slip between rows of grapevines. A Perkins diesel engine of 20 horsepower limits it to pulling a one-bottom plow.

Available in both gasoline and diesel models, the MF-135 fell in the 35-horsepower area. Both used a three-cylinder Perkins engine coupled to a 12-speed transmission. Gentner noted that there were some minor sheetmetal differences between the U.S. and U.K. models.

Introduced in 1964 and running through 1972, the MF-130 came with a four-cylinder Perkins diesel engine that tested in the mid-20-horsepower class. The transmission was an eight-speed. According to Gentner, the MF-130 was the French version of the MF-25, was produced only in France, and was upgraded to have the same sheet metal as the MF-135, MF-150, and MF-165.

"The MF-130 wasn't considered part of the 100 series line," Gentner explained. "It was really a series of its own. The 100 series was the MF-135, MF-150, MF-165, MF-175, and MF-180. The MF-130 fitted into the horsepower range that we needed here in the states. It replaced the MF-25 and was available in a high arch series with wide front axle."

From 1964 through 1975, the MF-150 replaced the MF-50. It was in the same power class as the

RED FOLEY SAVES MASSEY-FERGUSON?

It became obvious in the mid-1950s that Massey-Ferguson's marketing organization needed restructuring. Toward this end, in 1956, the decision was made to consolidate the North American tractor line of Massey-Harris of Canada and Massey-Ferguson of the United States into a one-line system. This eventually had a major positive impact on the distribution and marketing systems.

Ultimately the decision was made, and implemented, for the company to buy out existing dealerships and then establish new dealerships to handle the one-line products. As a result, the total number of dealers dropped from approximately 4,000 to fewer than 3,000.

The company then introduced national selling and advertising campaigns on a grander scale than ever before. This advertising blitz was coupled with an innovative and aggressive campaign to introduce new models to dealers.

In February 1959, all company dealers and their wives were flown to Detroit for the introduction of the MF-85. Dubbed the "Show of Progress," it reportedly involved the biggest one-day civilian airlift ever undertaken to date; and it necessitated the booking of 10 hotels and 3 banquet halls to accommodate the dealers.

The first-ever TV advertising of agricultural implements was a result of this event. In his book A Global Corporation, E. P. Neufeld said, "At the show the company introduced, with closed circuit television, a western music television program—the Red Foley group—which it then sponsored for a period on national television in the United States.

"On that program, farmers were told that they would receive a $100 check from Red Foley for every MF-35 tractor they purchased. No implement company had used national television or attempted such a promotion.

"It was an amazing success. Some still argue that Red Foley and not new management put Massey-Ferguson back on its feet in North America."

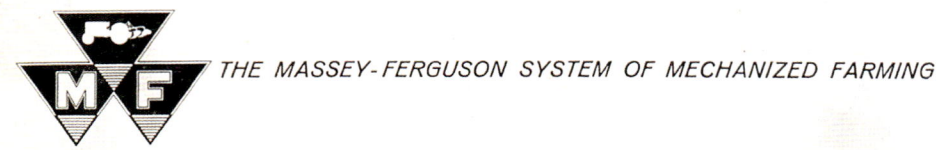

The cover page of the brochure for the 88 model touted it as "all new" and emphasized that the tractor possessed the lugging capability to get drawbar-attached implements through the field handily.

MF-135, but the MF-150 differed in being offered in different wheelbase and front axle packages. It featured the same engine and transmission as the MF-135.

The MF-150 had some options and variants, according to Gentner: "It featured different sheet metal and was designed so it would accommodate midmount cultivators. The MF-150, MF-165, and MF-130 had the high arch models. Both the MF-150 and MF-165 were available with a tricycle front end, as well as a wide front axle.

"The United Kingdom didn't manufacture the MF-150 but did make the MF-148, which was basically the same tractor, only for the foreign markets. The MF-148 wasn't imported for the U.S. market, although some did make their way in."

Replacing the Model 65 in 1964 was the MF-165. It stayed in the line for 10 years until 1975. Produced in both North America and the United Kingdom, the MF-165 was in the 45-horsepower range and available in both gasoline or diesel models. The diesel model used a four-cylinder Perkins engine, while the gasoline version featured a four-cylinder Continental engine. It came equipped with a 12-speed transmission. This tractor became the MF-168 when produced in the United Kingdom for foreign sales.

Also of 1965 vintage was the MF-175 manufactured in both North America and the United Kingdom. It fit into the 55-horsepower class and was powered by a four-cylinder Perkins diesel engine. It also used the 12-speed transmission.

U.K. tractors were built for the different markets on the continent and were diesel-only models; however, a small number of gasoline models were built at Detroit, Michigan.

The MF-180 was available in standard, high arch, or tricycle models and featured a raised operator's platform. It was only produced at the Detroit plant. Of 55-horsepower, it used a four-cylinder Perkins engine for the diesel version.

Massey-Ferguson's big tractor for the new line was the MF-1100. It developed close to 100 horsepower and featured the Category III hitch for big equipment. Power for the diesel model came from a six-cylinder Perkins engine.

It was also available in a gasoline model that tested out with just a few more horses. Fitted with the 12-speed transmission, this tractor was also "Detroit-only" production and stayed in the line through 1972.

MASSEY-FERGUSON 88

ENGINEERED FOR YEARS OF BIG-POWER PERFORMANCE

Brute strength lugging power is at your fingertips when you're at the convenient controls of the powerful Massey-Ferguson 88. With its rugged power, it really cuts wide-open spaces down to size. It easily handles heavy-duty operations with wide-level disc harrows, drills, tillers and chisel plows.

Check each of these features over carefully and read the facts behind it. In each you'll see the soundest engineering and dependability, the kind that promises years of low-cost, trouble-free, big-power performance.

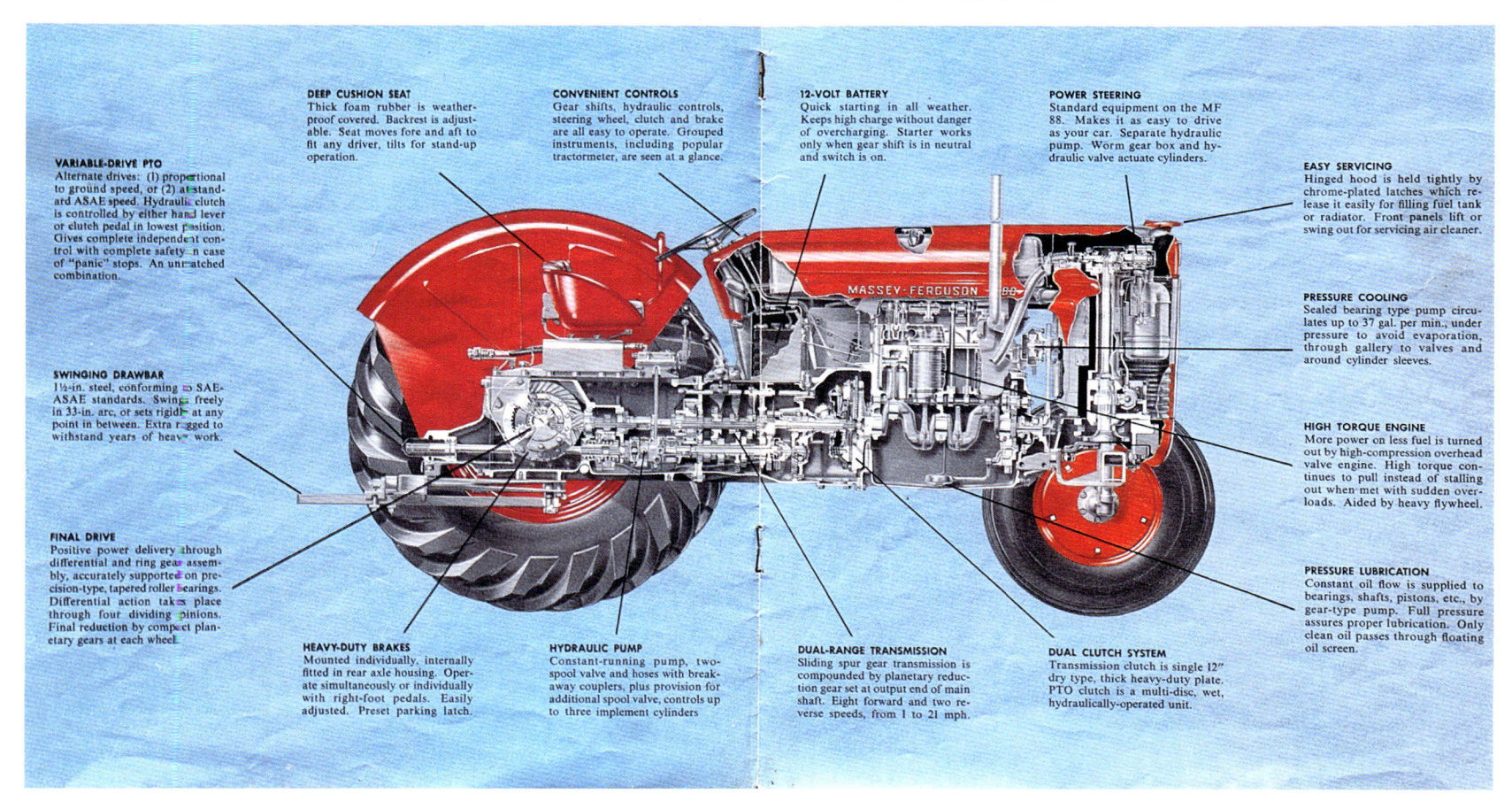

VARIABLE-DRIVE PTO Alternate drives: (1) proportional to ground speed, or (2) at standard ASAE speed. Hydraulic clutch is controlled by either hand lever or clutch pedal in lowest position. Gives complete independent control with complete safety in case of "panic" stops. An unmatched combination.

DEEP CUSHION SEAT Thick foam rubber is weatherproof covered. Backrest is adjustable. Seat moves fore and aft to fit any driver, tilts for stand-up operation.

CONVENIENT CONTROLS Gear shifts, hydraulic controls, steering wheel, clutch and brake are all easy to operate. Grouped instruments, including popular tractormeter, are seen at a glance.

12-VOLT BATTERY Quick starting in all weather. Keeps high charge without danger of overcharging. Starter works only when gear shift is in neutral and switch is on.

POWER STEERING Standard equipment on the MF 88. Makes it as easy to drive as your car. Separate hydraulic pump. Worm gear box and hydraulic valve actuate cylinders.

EASY SERVICING Hinged hood is held tightly by chrome-plated latches which release it easily for filling fuel tank or radiator. Front panels lift or swing out for servicing air cleaner.

PRESSURE COOLING Sealed bearing type pump circulates up to 37 gal. per min., under pressure to avoid evaporation, through gallery to valves and around cylinder sleeves.

SWINGING DRAWBAR 1½-in. steel, conforming to SAE-ASAE standards. Swings freely in 33-in. arc, or sets rigid at any point in between. Extra rugged to withstand years of heavy work.

HIGH TORQUE ENGINE More power on less fuel is turned out by high-compression overhead valve engine. High torque continues to pull instead of stalling out when met with sudden overloads. Aided by heavy flywheel.

FINAL DRIVE Positive power delivery through differential and ring gear assembly, accurately supported on precision-type, tapered roller bearings. Differential action takes place through four dividing pinions. Final reduction by compact planetary gears at each wheel.

PRESSURE LUBRICATION Constant oil flow is supplied to bearings, shafts, pistons, etc., by gear-type pump. Full pressure assures proper lubrication. Only clean oil passes through floating oil screen.

HEAVY-DUTY BRAKES Mounted individually, internally fitted in rear axle housing. Operate simultaneously or individually with right-foot pedals. Easily adjusted. Preset parking latch.

HYDRAULIC PUMP Constant-running pump, two-spool valve and hoses with break-away couplers, plus provision for additional spool valve, controls up to three implement cylinders

DUAL-RANGE TRANSMISSION Sliding spur gear transmission is compounded by planetary reduction gear set at output end of main shaft. Eight forward and two reverse speeds, from 1 to 21 mph.

DUAL CLUTCH SYSTEM Transmission clutch is single 12" dry type, thick heavy-duty plate. PTO clutch is a multi-disc, wet, hydraulically-operated unit.

Massey-Ferguson's MF-1130 was the company's first in-house–produced tractor to top the 100-horsepower mark. It was available in row-crop, standard, and Wheatland units. Offered as a diesel-only model, it was powered by a six-cylinder turbocharged Perkins engine coupled to a 12-speed transmission. The MF-1130 was in the line from 1965 to 1972.

Filling the 68-horsepower slot for Massey-Ferguson was the MF-1080. Powered by a four-cylinder Perkins engine, it was a "diesel-only" package with a 12-speed transmission. The MF-1080 was in the tractor line from 1967 to 1972.

Management at Massey-Ferguson had another awakening in 1965. Finally, the company came to realize that the biggest market for farm implements in North America was the Corn Belt in the United States. In fact, this area accounted for more than one-third of all farm equipment sales in the country. The Corn Belt also constituted the largest market for combines, balers, and implements and the second biggest for farm tractors.

Based on these facts, in 1965 Massey-Ferguson Inc. made a decision to move the company's U.S. executive headquarters and some manufacturing from Detroit, Michigan, to Des Moines, Iowa.

Another major change in the company's North American operations involved the sales and marketing arm of the company. There was previously only

When opened, the center spread of the 88 model brochure provided the prospective buyer with a cutaway side view of the tractor. Dependability was also indicated by "Engineered for years of big-power performance."

101

one general sales manager for the entire United States. This setup was replaced by a system of four divisional general sales managers.

Each division was given its own headquarters and staff and made into a profit center so that performance could be measured by profits and not by sales volume alone.

In 1966, the Industrial and Construction Machinery (ICM) group was established to help meet the rapid growth in that part of the company's product line. It would eventually be separated from the agricultural division and the Perkins engine operations in North America.

At this same time, the Landini facility in Italy was producing two models of crawler tractors for the ICM division.

JOINING THE AGCO FAMILY

By the early 1970s, Massey-Ferguson products were being built in 87 factories in 30 countries. Half of these countries were developing nations. Of its products, 93 percent were being sold in 190 countries outside of Canada, where it all began 123 years before.

The North American Massey-Ferguson distribution rights were purchased by the Allis-Gleaner

These were the five Massey-Ferguson tractors available when this brochure (you're looking at the back cover) appeared in the 1960s.

What's better in life than to have a great, growing family: corn knee-high on a successful diversified farm—and a new Model 180 to speed up chores or spread bulk fertilizer and do heavier work when needed?

102

Corporation (AGCO) in 1993. A year later AGCO purchased the worldwide holdings of Massey-Ferguson.

Here's how Gentner pointed out the ramifications: "Massey-Ferguson is a subsidiary of AGCO, and the name *Massey-Ferguson* was a worldwide name that was well accepted as a small- and medium-sized tractor. I'd say, all kidding aside, that Massey-Ferguson has kept the AGCO Corporation growing and advancing in the worldwide agricultural field.

"It was more or less a buyout. Victor A. Rice wanted to eliminate the agricultural products and go more into the industrial automotive field. It was more of a directional—personal interest—thing than a money issue that created the sale of Massey-Ferguson."

Regardless of the reason, or reasons, the legacy of Daniel Massey, Alanson Harris, and Harry Ferguson is still alive and well today at the Massey-Ferguson division of AGCO.

What started life as a 1964 Massey-Ferguson 2135 covered with industrial yellow paint has been restored to a 135, the comparable farm tractor model. Since it is an industrial unit, it features differential lock rear end and a foot throttle.

103

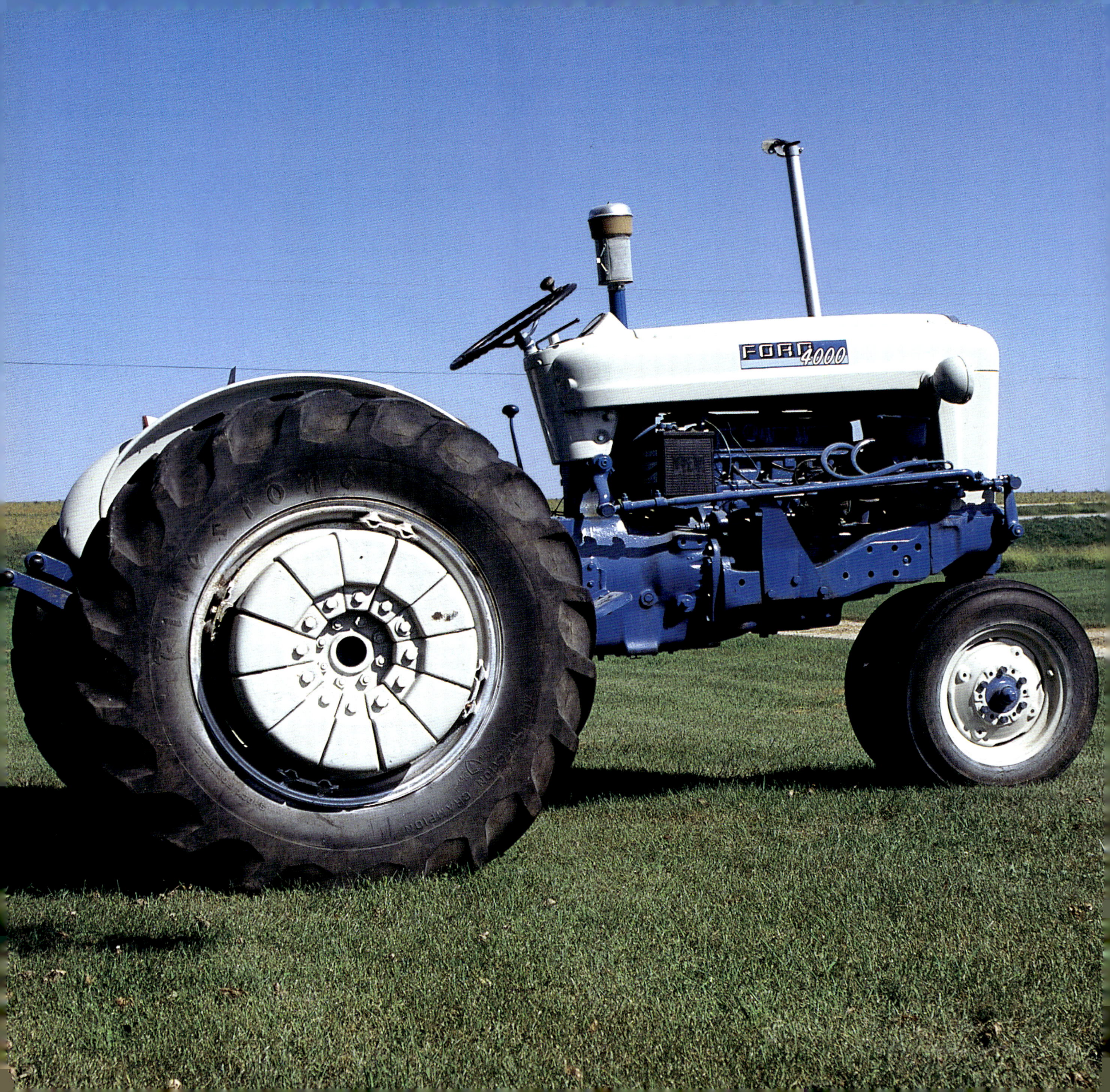
FORD
4000

CHAPTER SIX

FORD MOTOR COMPANY

Deere & Company and J. I. Case began producing automobiles after first establishing themselves in the agricultural implement business. Henry Ford reversed this process. Ford tractors arrived after Ford automobiles had already set the world on wheels.

The Ford Motor Company was established in June 1903, with 10 shareholders and a stock issue of $100,000; however, it barely escaped the same fate as Henry's two preceding company ventures, which ended in failure.

But escape and prosper it did, and three years later there were six dividend payments of $100,000 each. In a way, this could be viewed as seed money for later tractor development. The success of the Model T automobile provided Ford with adequate funds to continue his pet project of mechanizing America's farms.

The directors of Ford Motor Company didn't share Henry's enthusiasm for tractors. As a result, in 1915 he formed a new company, independent of Ford Motor Company, called Henry Ford and Son Inc., so he could pursue his tractor interests guided only by his whims and intuition. The company was formally sanctioned in 1917 with an issue of 10,000 shares, all held by members of Ford's immediate family.

The first of many experimental Ford tractors was completed in 1907. Time and resources always seemed to keep the project on the back burner, however, until the automobile production was well established.

Ford, by adroit maneuvering and intimidation, gained complete control of Ford Motor Company in 1919. By bringing both companies under his total control, there was no longer a need for Henry Ford and Son as a separate entity. It ceased to exist as a company in 1920.

Fordson production began in 1917, continued through three decades, and even today has prominent name recognition around the world. One event that helped elevate the name recognition was the outbreak of World War I.

With thousands of men and draft animals pressed into the war effort, food production plummeted. As the war wore on, England faced the prospect of being starved into surrender. Percival Perry, who later became Sir Perry, recognized the need for tractor power to increase and sustain food production. As it happened, he was also chief executive of the Ford Motor Company Ltd. in Great Britain.

Perry learned of the experimental Fordsons, then arranged for two of the prototype tractors to be shipped to the Royal Agricultural Society for evaluation in 1917. The results were affirmative for the Fordson. A license was granted for a new Ford factory to be built at Cork, Ireland, to produce the much-needed tractor. More immediate war concerns sidetracked the effort, however, when the facility was converted for aircraft production. Britain requested that Ford build the tractors in the United States and ship them over, which Ford did.

These efforts gave the Ford Fordson a strong market foothold in Great Britain, one that IHC and other competing tractor companies resented. By 1928, more than a half-million Fordson Model F tractors found their way down the assembly line and into barnyards and fields of farmers around the world.

This 1964 Ford 4000 is virtually identical to the 1962 961 model, with the exception of a few minor "modernization" items and the switch in paint colors. The engine produces 41 horsepower.

105

Purchased in early 1960, this Ford features the Elinco front-wheel assist made by Elwood Engineering Company. The tractor operates in four-wheel drive at all times, although later models were able to switch to either two- or four-wheel drive. This particular tractor also features a Sherman over-and-under auxiliary transmission with 12 speeds forward and 3 speeds reverse.

When Ford withdrew the Fordson from U.S. production in 1928, it wasn't the end of the Ford tractor. Ford had always stated a desire to provide some enterprise in his native Ireland. So he packed up the Fordson tooling and equipment from the Rouge plant and shipped it to Cork, Ireland. Production got under way early in 1929 to meet the demand for spare parts and tractors for Great Britain and other export markets.

Several improvements, most of them to the engine, were made. The new U.K. Fordson was called the Model N. Some of these Irish models were even exported to the United States by the Sherman and Shepherd Company of New York.

Three years later, the tooling and equipment were once again packed up and moved, this time to Dagenham, England. Economics dictated the move since tractor sales had slumped to a level that made it impractical to continue production at Cork.

In 1937, the Fordson All-Around provided the company with its first row-crop tractor that included a better PTO design. In order to keep up with the new bright paint schemes that were being sported by American tractor manufacturers, the Fordson's color was changed from dark blue to bright orange.

The Fordson tractor had sold in amazing numbers around the world. Yet Henry Ford didn't feel that it had achieved his goal of really mechanizing farming. While his Fordson was being manufactured at the Dagenham facility, Ford was funding experimental work at Dearborn. A wide and sometimes strange array of experimental tractors was made and tested on the Ford experimental farm.

Row-crop designs powered by six-cylinder and V-8 engines joined the fleet of experimental tractors. These were the more traditional designs, but Ford liked to push the design envelope with prototypes even if they only proved what wouldn't work.

It took another Irishman to present Ford with the right design at the right time—and to embody Ford's concept of the ideal tractor. Eccentric, dogmatic, but brilliant, Harry Ferguson brought his Ferguson-Brown tractor and three-point hitch system to Fair Lane, Michigan, in 1938.

The famous handshake agreement was reached. Once again Ford was in the tractor business. This verbal agreement gave Ford the responsibility of manufacturing the tractor, while the

Starting life as the two- or three-plow 671, in late 1962 Ford painted it blue and gray and called it the Ford 2000. This 1961 model has the Select-O-Speed transmission. Featured on production models beginning in 1959, the Select-O-Speed transmission was described as a "full power shift" with 10 forward speeds and 2 reverse. Early Select-O-Speed transmissions were troublesome and many failed after only a short time. After extensive redesign, the problems were corrected.

Known as the 541 Offset Workmaster, this 1962 tractor is, in essence, a take-off of the 701 model. Ford made many modifications to obtain the offset features in an attempt to obtain some of the Farmall A's market share. The 541 was used mostly by truck farmers to cultivate lettuce, tomatoes, and tobacco. The 541 came with 28-inch tires and the 38-inch-equipped model was designated the 541-4.

"Rode hard and put away wet" might describe this 961. It's been laboring hard ever since it left the factory in 1962, yet is still being used daily for farm chores and light fieldwork.

sales and distribution of tractors and implements fell to Ferguson.

After the launch and successful acceptance of the new N Series tractor, Harry Ferguson wanted the British Ford Company to replace production of the Fordson with the new tractor. For various reasons, even despite Henry Ford's influence in favor of Ferguson, the British plant continued to produce the Fordson.

Henry Ford died in 1947 at the age of 83. Although he's most remembered for his stunning success with the Model T automobile, at the time of his death there were 1,700,000 reasons why he should also be remembered for his contribution to agriculture: the number of Fordson and Ford tractors that had been produced in the three decades since the introduction of his Fordson tractor.

In 1946, only a few months earlier, the handshake agreement had run its course. Ferguson and his marketing system were scuttled and replaced with a Ford counterpart called Dearborn Motors.

FORD DIESEL TRACTORS

...the lowest priced diesel tractors in their class

Here are some of the outstanding features of Ford's new diesel tractors

3-4 PLOW POWER—There's plenty of power to keep going in the hard pulls because engine torque stays up when speed is pulled down.

SIMPLE STARTING—Starts directly on diesel fuel.

EFFICIENT COMBUSTION—Ford's system of direct injection results in maximum power from minimum fuel.

EASY MAINTENANCE—Simple rugged design provides convenient, low cost maintenance.

And many more features that mean you get more, save more with a Ford Diesel tractor.

For years, Ford tractors have been helping farmers do their work better, easier and at lower cost. They've earned a well deserved reputation for versatility and general all 'round usefulness that contribute to more profitable farming.

And *now* it's the *Ford Diesel*, a diesel tractor that's practical for the average farmer! The cost differential of a Ford Diesel Tractor over that of a comparable gasoline tractor is much *less* than that of other diesels over gasoline models. This, with its low operating costs means it can return a profit on fewer total hours of use.

You can save plenty with a Ford Diesel Tractor because it's priced to help you *profit more*—on more farm jobs!

AVAILABLE IN ALL 801 AND 901 MODELS

● **SPECIAL UTILITY**
Model 821 Diesel

Ford's famous dependable power at the lowest possible cost—that's what you get in the model 821 diesel. Suited for pull-type equipment, this special utility model is designed for those jobs where tractor hydraulic system and 3-point hitch are not required.

● **ALL-PURPOSE**
Models 841, 851, 861 Diesels

Ford diesel power brings new operating economy to these 4-wheel all-purpose tractors that have built a reputation on low cost performance. They're compact and maneuverable, adaptable to just about any type of job. Extra low center of gravity means good stability on uneven terrain.

● **ROW CROP**
Models 941, 951, 961 Diesels

Don't overlook the savings you can make by putting Ford diesel power to work for you on your row crop work. Ford row crop diesels are built to make row crop farming easier, more efficient, more economical. Available with your choice of front ends. Power steering and power adjusted rear wheels are standard equipment.

Single Front Wheel · Wide Adjustable Front Axle · Dual Front Wheels

---

Mounting a unique one-arm Model 19-20 loader that was designed to be quick-on and quick-off made this 1960 971 LP gas model really handy around the farm. Its Select-O-Speed Power-Shift transmission with 10 speeds directs 50 horsepower from a 172-cubic inch engine.

Ford brochure stresses economy and value for dollar invested from Ford diesel tractors, pointing out that they're "... the lowest-priced diesel tractors in their class." It wasn't until 1948 that Ford offered a diesel-powered tractor when it began offering a Perkins diesel engine in the E27N tractor. In 1952, the New Major from the Dagenham plant was the first Ford tractor to use a Ford-built diesel engine.

109

Beginning in 1964, a slightly different shade of gray and blue became the worldwide Ford tractor colors—supposedly because they didn't fade as readily. This 1964 2000 offered 39 horsepower in addition to a new look in paint style.

The last Ford-built Ford-Ferguson was delivered to Ferguson's organization in July 1947. After that, Ford marketed a new tractor, the Ford 8N, which supplied some badly needed new features. The success of the new 8N tractor and implement line was unquestioned. In both 1948 and 1949, production surpassed 100,000 units per year.

Still, Ford offered just one tractor model, the 8N, or two models if you include the E27N, which was still in production at the Dagenham plant. The E27N had an improved final drive and a three-point hitch, but the engine was essentially the same as the previous model.

In 1948, the Perkins P6 diesel was announced as an option on the E27N, giving Ford its first diesel-powered tractor; however, the diesel concept had been championed by Patrick Hennessy of Ford UK since 1944, when he pushed engineering to design a Ford-built diesel engine.

Production models of the new diesel appeared in the New Major from Dagenham in 1952. The same

What's in a name? Workmaster certainly gave the impression that this new lineup of Ford tractors could get the job done. Note that three front ends were available on row-crop models: narrow front single wheel, narrow front dual wheels, and wide adjustable front axle. The other basic style was the "All Purpose."

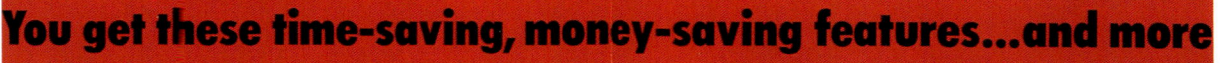

This 1967 6000 Commander LP is a rare model, because so few LP gas models were ever delivered, especially to the Midwest. This one receives 66-PTO horsepower from a 222-cubic inch engine. The Model 6000 was announced in the fall of 1961 by Ford Tractor Operations. It was available with gasoline, diesel, or LP gas Ford-built six-cylinder engine. The Model 6000 had many advanced features such as the Select-O-Speed transmission and an accumulative hydraulic system. Probably rushed into production, many of the new features failed to live up to expectations and the company eventually had to replace the entire tractor for customers who purchased the first production run.

111

FORD 3000 VINEYARD

Only 44.5-inch overall width with full three-plow power!

Here is full scale tractor power in a popular horsepower range with 30- to 52-inch rear wheel tread. At the narrow tread setting, hub to hub width is only 44.5 inches. This compact design will enable many vineyard owners and growers of other crops to utilize closer row spacings to increase yields and income from their land.

- **Husky diesel engine** delivers 39.2 PTO horsepower.
- **Complete Ford hydraulic system** with draft, position and flow control. Category I three-point hitch.
- **Transmissions:** Eight-speed manual shift with "live" power take-off or ten-speed power shift with independent power take-off.
- **Differential lock** and horizontal exhaust are standard equipment.

Probably the narrowest in width of any tractor of its power, the 300 Vineyard model was designed to easily work between traditionally close-set grapevines. Yet the tractor could still pull a three-bottom plow, if necessary.

engine was also available in gasoline and gasoline-kerosene versions.

In 1953, Henry Ford II paid homage to his grandfather's dream of revolutionizing farming by introducing the Ford NAA Golden Jubilee to mark the 50th anniversary of Ford Motor Company. Isn't it interesting that a tractor, not an automobile, was the product chosen to celebrate this noteworthy company milestone?

The NAA Golden Jubilee was produced for one year only, 1953. It introduced Ford's new Red Tiger overhead-valve, four-cylinder, 134-cubic inch engine rated at 30 horsepower. This "new" engine was really Ford's overhead-valve six-cylinder engine shortened by two cylinders. Live PTO was offered as a dealer-installed option. The most significant change was styling, and, oh yes, the hydraulic system was re-engineered to circumvent the Ferguson patent.

Dearborn Motors became wholly owned by Ford Motor Company in May of the Jubilee year. By August, Dearborn was replaced by the Ford tractor division operating out of the Birmingham, Michigan, offices. This arrangement was altered again in early 1954 when the Ford tractor division was restructured as the tractor and implement division.

Buyers finally had a choice with the 1955 models, as Ford abandoned its "one tractor fits all" philosophy, which it retained for 40 years. The company introduced five new tractors in the 600 and 800 series. The Model 600s used the same Red Tiger engines as the Jubilee. The four-speed transmission was standard in the 640, while the 650 and 850 introduced a new five-speed tranny.

Farmers who wanted the five-speed transmission plus a live PTO could opt for the 660 or 860 models. The 800 series offered three-plow performance. The 850 and 860 used a bored-out Red Tiger engine of 172 cubic inches to generate an additional 10 horsepower. Both models used the five-speed transmission.

The first tricycle models were the 700 and 900 series. The smaller 740 used the 134-cubic inch engine, while the 950 and 960 had the larger 172-cubic inch powerplant.

The Ford Workmaster and Powermaster came out in 1957. The Workmaster series began with the 601, and the Powermaster started with the 801 model.

Ford's 801 Diesel was the company's first tractor to use the U.S.-built Ford diesel tractor engine that appeared in the line in 1958. For the first time Ford could offer the same three engine fuel options as most of its competition: gasoline, diesel, and LP gas.

Ford stepped up its efforts to be a major player in the tractor business. The Highland Park, Michigan, plant underwent a $6.7 million renovation program in 1959. By 1960, Ford's line boasted a total of 32 models, including the U.K. Fordson Dexta and Major.

Towing aircraft around airfields was the job of this 1964 Military Tug. Similar to a standard 2000 series model, it differed in having a special drawbar, special transmission, a horn, four-way flashers, block heater, cab, and a heavy-duty alternator instead of a generator.

Rather than mounting the usual generator, the 1964 Military Tug substituted a heavy-duty alternator.

113

TWO FORD ORPHANS

Ed Pinardi described the early years of the tractor engineering department this way: "We were like orphans. Nobody wanted us because we were so small, as we were building only the 8N tractors at that time. Then finally we got our own building and own division when Dearborn was bought out.

"After Ford Motor Company bought out Dearborn Motors we moved up to Birmingham, Michigan, and that's where we settled down and became the Tractor Engineering Center.

"Engineering was all one division together, Lincoln engineering, truck engineering, car engineering, all were together. We were just a small division and we didn't fit in, more or less for better or for worse, with the rest of the automobile industry," he remembered.

"We were in the same building on the same floor. I could walk to Lincoln Mercury, I could walk into electrical engineering, I could walk into the engine build-up and test. We were just one big building on one big floor."

The orphan analogy is one that Pinardi can personally relate to as he explained, "I started with Ford in 1937. I was an 'orphan' too. I was the 13th child in my family and considered an orphan.

"At that time Ford Motor Company was giving orphans who didn't have money to go to school a chance to succeed in life. I was chosen as one of them to come to Michigan to go to work for Ford. I started in 1937 right out of high school.

"I went into tool and die and became a journeyman. Then I enlisted in the U.S. Navy and went to war. When I came back Ford Motor Company put me through engineering school. I retired after 48 years."

"It's a dirty job, . . ." and certainly one that nobody ever volunteered to do. Yet, routinely hauling out and spreading manure was important both to keep pens clean and to lessen the need for commercial fertilizer. Driving this 601 probably made the task slightly more pleasant.

FORD TRACTORS IN THE 1960S

There were no new tractors in the Ford stable when the 1960s began. One of the models that made the decade transition was the 1959 871, painted gold and featuring Ford's new Select-O-Speed, shift-on-the-go, 10-speed transmission. Perhaps as few as 100 of these gold models were made for distributors' showrooms and demonstrations.

Not many, if any, were trouble-free and fewer still exist today. According to retired Ford implement dealer, Don Horner, "Many of the first Select-O-Speed transmissions have had stick transmissions installed."

In retrospect it would be a sure bet that Ford management and engineers wished the standard transmissions would have stayed around a bit longer.

Edward Pinardi, who retired as senior product design checker at Ford, had a long career as an engineer at Ford's tractor division. This is how Pinardi remembered the problem, noting, "At the time of the 10-speed problems the tractor division shut down almost completely and had only limited production of some models. All of the early production transmissions were replaced and brought back to the factory for analysis.

"Engineering did continue with new model design; however, most of the department's time was devoted to repairs and problems concerning the new transmission. When we put it out in the field some lasted 50 hours, others 20 hours."

Pinardi adds that, "Harold Brock, who headed Ford's tractor engineering, said it wasn't ready, and that the transmission was far from being ready to put into production.

"Then what happened was that he went on vacation. While Brock was gone from the office some of his bosses came in and said it was ready and to proceed. Marketing said it was ready. So we put it out and that's when we ran into trouble, serious trouble."

There were even more woes awaiting the Ford tractor operations when the 6000 was rolled out in 1961. The 6000 was a big tractor loaded with features. It was equally loaded with problems.

Power came from the same basic 223-cubic inch, six-cylinder engine design found in Ford trucks. It was available in the 6000 tractor with gasoline, diesel, or LP gas engines. The gasoline and LP gas powerplants were 223-cubic inch engines and the diesel was a 242-cubic inch engine. PTO horsepower was 66.2 and 66.8, respectively.

Customers had a choice of tricycle row-crop or wide front axle. The transmission was the 10-speed Select-O-Speed. Power steering and power disc brakes were standard equipment, as was lower-link sensing on the three-point hydraulic system.

Don Horner has a lifetime of experience with Ford tractors. His father owned the Ford dealership in Geneva, Ohio, and Horner continued to operate the dealership until his own retirement. Here's how he viewed some of the Ford tractors of the 1960s:

"The Select-O-Speed was introduced in the late 1950s and was so bad that Ford gave everybody who bought the initial shipment of tractors a new transmission. This transmission came out on the 801 series, but could also be had on the 601.

"It was the only transmission Ford ever put in the 6000, an advanced type of tractor. When Ford introduced the new World Tractor the 6000 became the Commander 6000. There was a color change too. The first 6000s were so bad that Ford gave everybody a whole new tractor except for the wheels and tires.

Every dealer received a gold dealer demo 871-D Powermaster to demonstrate the shift-on-the-go feature known as Select-O-Speed. The gold paint made it stand out from all other tractors. This particular tractor is extremely rare. It has 540 PTO, power steering, 172-cubic inch diesel engine, and power-adjust rear rims.

115

NEW FORD 4000/5000 Orchard Specials

Ford 5000 Orchard Special offers big 67* PTO hp in a sleek low-silhouette tractor. The 52* hp Ford is similarly designed.

Now, Ford Blue-Key quality can be yours in a sleek, low orchard tractor. Get all the rugged strength and power built into 52* PTO hp Ford 4000 or 67* PTO hp Ford 5000 All-Purpose models. Low silhouette permits working under trees where ordinary tractors can't go. Clean lines reduce chances of snagging branches; make your orchard jobs go faster and easier.

Ford has reduced overall height with lower front axles and smaller diameter wheels. Air-cleaner intake is recessed. Headlights are mounted in the grille. Rear fenders and steering wheel are lowered about two inches, compared to all-purpose models. It all adds up to as much as five inches less height at the hood; a rear silhouette lowered as much as nine inches; depending on tire options chosen.

Either tractor is available with Ford Select-O-

Speed. This time-proved 10-speed, power-shift transmission is unsurpassed for PTO work. You get the efficiency of an all-gear drive, plus shift on-the-go convenience. Fully independent PTO is power-engaged, with never a jerk or clash of gears.

Or you can order your Ford 4000 or 5000 Orchard Special with simple and rugged 8-speed transmission, also available with independent PTO.

Engines, transmission, hydraulics and final drives are the same that make Ford 4000 and 5000 tractors best choices in All-Purpose models. Orchard models offer the same comfortable operator's area; the same convenient controls.

Ford Blue is the growing thing. Compare. You'll agree Ford is your best choice.

*Maximum observed PTO hp, 8-speed transmission.

A low silhouette was important to 4000/5000 Orchard Special tractors so they could be driven under trees where bigger, taller tractors couldn't readily work.

"I went to the introduction in Columbus, Ohio," he continued. The tractors were sitting on the ramp outside the auditorium just leaking oil on the ground like crazy. The design was really advanced, but it was too hurried into production.

"It used a closed hydraulic system with a nitrogen accumulator. Commercial airplanes also have an accumulator—essentially a big tube full of nitrogen gas—that accumulates hydraulic power. In an airplane when the pilot lowers the landing gear there's a sudden burst of hydraulic power due to the accumulator," he said.

"The tractor only had a 4-gallon-per-minute pump, though. So the advantage is when you are going down the field this small pump is pumping up the accumulator.

"It used to surprise the farmers at the fair when they'd see that even after the demo tractor had its engine shut down the plow could still be raised. They weren't used to anything like that. The 6000 was strictly American-made, and it had all kinds of problems."

The Workmaster series was renamed the 2000 series in 1962 at the same time the Powermaster series was designated as the 4000 series. These U.S. tractors carried the new blue-and-gray, really closer to white, color scheme, as did the imported Fordson Super Dexta and Super Major. U.K. diesel Super Dextas were sold in the United States as the 3000 and the Super Major as the 5000.

Horner pointed out the different transmissions Ford offered, saying, "The Select-O-Speed was carried through, with improvements, on into the World Tractor introduction in 1964 and then on until some time in the 1970s.

"The Select-O-Speed was never an eight-speed model. We didn't have an eight-speed until we went with the unified line in 1964. The eight-speed was strictly a dual range—a four-speed with a hi-lo.

"An interesting point," he added, "is that the first eight-speed that was introduced in 1964 really only had seven speeds. That's because fourth gear was the top of the low range and fifth gear was the bottom of the high range—they were both the same numerical range. I don't know why Ford did this, but it was changed later on."

From this time forward all Ford tractors, regardless of where they were manufactured, became part of the company's "World Tractor" program. The Ford World Tractor concept addressed the same problem that other major tractor manufacturers were also facing.

As tractor models proliferated, companies with manufacturing plants located in several different countries had to standardize designs and reach as much commonality as possible in parts and components in order to control production costs.

Horner said that the theme of the World Tractor's New York City introduction was, "It's just like it is back home."

This is reflected, as Horner showed, in the transmission offerings: "Both the Dexta and the Major had six-speeds until 1964. After that, by being part of the World Tractor line, they could have a four-speed or eight-speed manual transmission or the ten-speed Select-O-Speed. The farmer had a choice of these three transmissions after the World Tractor was introduced."

The year the World Tractor was introduced, 1964, was the same year that U.K. tractor production was moved again, this time from Dagenham, England, to a new plant 20 miles away in Basildon.

This brought the 1000 Series on-line with tractors being built in both the United States and the United Kingdom. Production in the United States remained at Highland Park, Michigan. Meanwhile, the tractor division had switched Ford's European facilities at Antwerp, Belgium, to tractor production.

"Most of the components for the World Tractor were made in Dagenham, England, and assembled in Detroit," according to Horner. "We're talking about major components, too, transmissions, engines, and final drives. Other parts were procured locally in whatever country the tractor was being assembled.

"As I understand it, two years before the introduction of the World Tractor, components had been arriving at the Highland Park plant. There was a mixed-up mess when assembly began.

"While a lot of this stuff would fit together, because the parts were interchangeable, they still might not be correct. As a result, some tractors were assembled that really didn't match specs for any model.

"Ford had to send service people up from the Columbus, Ohio, factory branch to help straighten it out so the right components ended up in the right model," he said.

In 1968, the Dearborn-built 8000 gave Ford its first tractor that had to use three digits to show its horsepower rating. The 8000 checked in at 105-PTO horsepower. A year later the 9000 model upped the ante to 130 turbocharged horses.

Pinardi commented on other engineering developments in this time period: "Was Ford making its own engines? Well, yes and no. Ford was improving on the engines it had before while trying to get a place to build its own engines. The engines were being built by another division of Ford Motor Company for the tractor division.

"The tractor division eventually did get its own new engine facilities in Romeo, Michigan, 30 miles away from the tractor engineering plant in Birmingham, Michigan, in the late 1960s or early 1970s; however, during the 1960s tractor production was at Highland Park.

"We were starting in the 1960s to develop a tractor cab," he pointed out. "Of course, we went

A 3000 tractor combined the attributes of enough power to do the jobs for which it was designed and agility. Here one is spreading fertilizer on individual trees while driving down the rows of an orchard.

HANDY, ECONOMICAL POWER FOR BIG CAPACITY PERFORMANCE!

There's a model tailored to your needs

Yes, whatever your type of farming, there's a tractor in the Ford *Powermaster* series to meet your 3-4 plow power requirements. All models are unusually versatile, and are adaptable to a wide range of profitable uses. Pick the one that suits you and your needs, and you'll find that it's "tops" for dependable performance!

• ALL PURPOSE MODELS

These 4-wheel tractors pack a lot of power for versatile performance. They're unusually well suited for use with all types of rear attached and pull type implements, harvesting machinery, etc. Low center of gravity and compact design result in stability and maneuverability for dependable performance characteristics in all types of terrain.

ROW CROP MODELS •

Here's big-capacity, easily maneuvered power for all types of exacting row crop work. Provides for quick, easy mounting of 2- and 4-row front mounted cultivating, planting and fertilizing equipment. Also handles all types of rear mounted and pull type equipment. Front end options and rear wheel spacings for practically all crop conditions. *Power steering and power adjusted rear wheels* are standard equipment on all Powermaster row crop models.

DUAL FRONT WHEELS
SINGLE FRONT WHEEL
WIDE ADJUSTABLE FRONT AXLE

...*plus* special utility models

Maximum drawbar power at minimum cost... that's what you get in the special utility models! They have the same Powermaster engine and many of the other features of the All-Purpose models, but eliminates the cost of deluxe equipment not needed for drawbar work. P.T.O. is optional.

COPYRIGHT 1957, FORD MOTOR COMPANY • DEARBORN, MICHIGAN

Here are some of the things you get in a FORD POWERMASTER TRACTOR

Powerful "RED TIGER" engine

There's a lot of efficiency and economy built into this big Ford Red Tiger engine. Its big diameter bore and short stroke pack a powerful punch ... keep friction and wear to a minimum. Lugging power helps to pull through tough spots without stalling. Develops approx. 50.0 belt horsepower (manufacturer's rating for gasoline model).

- **3-POINT HITCH** for quick, easy mounting of rear attached implements ... saves time, labor and money. Linkage, together with the hydraulic system provides for automatic weight transfer for heavy draft implements.
- **BIG CAPACITY HYDRAULIC SYSTEM** provides plenty of live hydraulic power for instant, positive control of 3-point hitch and remote hydraulic cylinders.
- **PTO—LIVE OR STANDARD** depending upon model.
- Efficient power for operating PTO machines such as combines, balers, corn pickers, forage harvesters, etc.
- **SWINGING DRAWBAR*** provides for efficient handling and control of pull type equipment.
- **COMFORT AND CONVENIENCE** to make operation easier, less tiring. Everything from deluxe Rest-O-Ride* seat to position and angle of steering wheel have been designed with operator in mind.

*Model 821 equipped with fixed drawbar and standard seat.

The Powermaster tractor mounted a "Red Tiger" engine to match the red belly paint scheme. It was advertised as having excellent lugging power for the tough spots. The Red Tiger engine debuted with the 1953 Golden Jubilee model. This "new" engine was Ford's overhead-valve six-cylinder with two cylinders lopped off.

whole hog in developing an elite cab. The 1960s were primarily the time of R&D as far as cabs were concerned."

Later the cab was referred to as the "million dollar cab," because that was the price tag on the tooling.

The first true Ford industrial tractor models had appeared in the line in 1958. The 1801 series featured a longer wheelbase than the agricultural models and had a full-length subframe on which to mount the loader and backhoe. These first industrial units featured the familiar red-and-gray paint.

The 2000 and 4000 utility, or light industrial units, had two notable differences from the agricultural versions. The front axle was one piece and the differential was much stronger. The paint scheme was evidently under transition; however, these industrial tractors were painted blue and yellow.

Under the World Tractor program, the 3400, 4400, 3500, and 4500 were introduced as industrial versions in 1965. The 3400 and 4400 acquired the nickname "soft nose" for the standard agricultural sheet metal around their radiators.

The 3500 and 4500 were called the "hard nose" models due to a heavy-duty radiator cover that protected against damage on an industrial job site. By now the industrial line featured all-yellow paint jobs.

To sum up the decade, Pinardi approached it this way: "The 1960s was a time like the changeover from the Model T to the Model A at Ford. For us, it was a long-term, continuous engineering effort. We

118

The smaller models of the Ford tractors made during the 1960s weren't limited to solely working on farms. Others in various industries, such as golf course greenskeepers, prized the tractors for their nimbleness and ease of operation.

were continually trying to advance in horsepower and bigger tractor designs."

A GREAT NAME DISAPPEARS

As the power race escalated, Ford really jumped in the fray during the 1970s with models that pushed Ford's top horsepower up to 186. Needing still more horses, Ford cut a deal with Steiger Tractor Company of Fargo, North Dakota, to put their megatractors in the line complete with Ford colors and Ford emblems. At last Ford could supply buyers with models all the way up to a whopping 300 horsepower.

Then, in a move to round out its line of agricultural equipment, in 1985 Ford bought Sperry New Holland. This brought into the Ford fold a leading manufacturer of harvesting machinery. It also caused a name change to Ford–New Holland.

Case IH purchased the Steiger company in 1986. This left Ford searching for a replacement for these top-of-the-line-horsepower tractors. Within a few months, Ford bought the Versatile line of high horsepower, articulated four-wheel-drive tractors.

The Ford tractor and agricultural division was altered once again in 1990 when Fiat became part of the mix by integrating Ford–New Holland and FiatGeotech into one of the world's largest tractor manufacturers. Fiat, headquartered in London, ended up with an 80 percent interest in the combined companies. The Ford facility at Basildon, England, served as the tractor and engine plant for the new enterprise.

It's no secret that Henry Ford liked to maintain absolute control over his business empire. It's doubtful that he would have condoned the merger that removed the Ford name from agricultural tractors.

On the other hand, if he could see where his Fordson has led he almost surely would be both impressed and pleased.

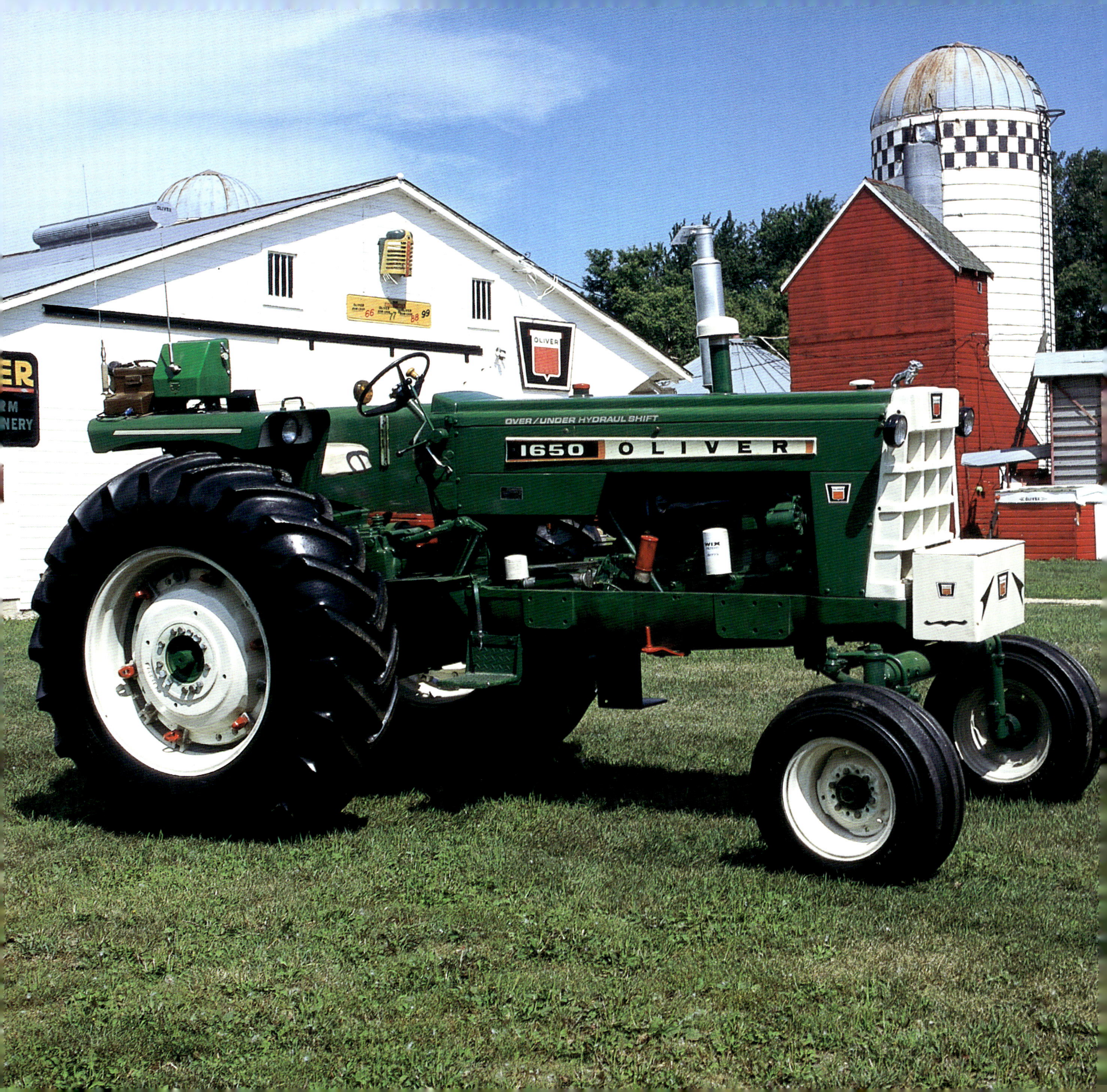

OVER/UNDER HYDRAUL SHIFT
1650
OLIVER

CHAPTER SEVEN

WHITE FARM EQUIPMENT COMPANY

OLIVER

Born in Scotland on August 28, 1823, James D. Oliver immigrated with his family to the United States when he was 11 years old. In 1855, Oliver bought a quarter interest in the South Bend Iron Works, a foundry in South Bend, Indiana. This venture led to the creation of the famous Oliver "chilled plow."

Oliver perfected the process of chilling cast iron, which allowed it to take a high polish with adequate strength for use in plowshares and moldboards. Like John Deere's steel plow, Oliver's chilled plow provided farmers with a tool that would open the vast acres of midwestern prairie. The plow also gave Oliver the foundational piece of agricultural equipment upon which he built an empire.

In 1868, the foundry incorporated as the Oliver Chilled Plow Works. Oliver died in 1908 and his son, Joseph D., then became CEO.

In 1929, Oliver Chilled Plow Works joined a merger that included the Hart-Parr Tractor Company of Charles City, Iowa, and Nichols and Shepard Threshing Machine Company of Battle Creek, Michigan. The American Seeding Machine Company of Springfield, Ohio, joined the group one month later. These four companies formed the Oliver Farm Equipment Company.

Although the Oliver company was developing its own tractor before 1929, the roots of the Oliver tractor can be attributed to Charles Hart and Charles Parr. The two men met at the University of Wisconsin in Madison in 1894 and established the Hart-Parr Company in Madison in 1897. The following year, they were already building stationary engines. (An interesting feature of these early engines, carried over to their early tractor motors, was the use of oil as an engine coolant. This eliminated the possibility of the coolant freezing during cold weather and damaging the engine.)

In 1901, they moved the business to Charles City, Iowa, in search of better financing. Here they produced the initial Hart-Parr gasoline traction engine—the Hart-Parr No. 1—built in 1902 and sold to an Iowa farmer. The Hart-Parr No. 2 was also a 1902 product, but it didn't sell until 1903. By then the company was producing new models, the 17-30 and 22-40.

In the years before the Oliver merger, the company introduced numerous other successful models and features. The company was the first to mass-produce an internal combustion tractor and the first U.S. manufacturer to have substantial foreign sales.

According to T. Herbert Morrell, author of *Oliver Farm Tractors*, "The independent power take-off (PTO) was first introduced on the Hart-Parr 18-36 in 1928." This is 20 years prior to the year in which Cockshutt of Canada is sometimes incorrectly credited with featuring the innovation.

After the two companies joined ranks, the combined tractor line typically included both names,

Thanks to a high-speed rear end, this 1968 1650 "Really goes down the road!" according to its owner, at nearly 30 miles an hour. It has over/under hydraulic shift and three-way Hydra-Power torque. Horsepower was 70. Nebraska test results No. 873 showed the 1650 diesel producing 66.28-PTO horsepower, while the gasoline version tested at 66.72-PTO horsepower.

121

Although built in 1960, this 990 is still being used for fieldwork. Rated at 80 horsepower, the owner reports this particular tractor has developed 105 horsepower on a dynamometer test. Its engine is a three-cylinder diesel made by GMC. Oliver made this same tractor for Massey-Ferguson as the 98 model. The only differences were the grille and paint color.

"Oliver/Hart-Parr," from 1930 to 1939. Then Hart-Parr was dropped from the name. During the 1930s, the company offered standard, industrial, row-crop (with single and dual front wheels), orchard, and airport models. In 1935, the company offered a six-cylinder engine and electric lights as options.

In 1940, Oliver sought the services of industrial designer Wilbur Henry Adams to style certain models. It began research and design efforts for the outstanding Fleetline models the same year. The Fleetline series had ambitious objectives, including improved appearance, different fuel options, practical independent PTO, maximum interchangeability of parts, and operator well-being.

The new Ridemaster seat, designed around torsional rubber springs, was introduced on the Fleetline models in 1949. They gave Oliver tractor operators one of the best rides in the industry.

Oliver bought the Cleveland Tractor Company of Cleveland, Ohio, in 1944 to strengthen its industrial line. That same year the company's name was changed from Oliver Farm Equipment to the Oliver Corporation.

Oliver, similar to its competition, was bitten by the expansion bug. It continued to acquire several companies until it, in turn, was bought by White. Oliver purchased such companies as Be-Ge, a hydraulic equipment manufacturer located in Gilroy, California; Farquhar-Iron Age of York, Pennsylvania, manufacturers of potato planting and harvesting equipment plus conveyors; and Chris Craft Outboard Motors.

In the 1950s, Oliver introduced the Super line in an effort to increase the power of its Models 66, 77, and 88 and stay in the industry's horsepower race. Styling remained pretty much the same, although the side panels on the engine compartment covered less of the engine in order to provide better air circulation. Also during this period, the company began working with industrial designer Wally Droegemueller, whose styling influence continued into the White tractor era.

WHEN FARMERS GOT TO CHOOSE TRACTOR COLORS

Oliver's exhibit at the biggest state and sectional fairs in 1937 featured a voting table in the midst of six row-crop 70 tractors painted in special color combinations.

After looking over the tractors, farmers then indicated their preferences on ballots. As an incentive, each farmer who voted received a leather pocket key case.

The colors were: Chrome Green body, red trim, ivory lettering; Regatta Red body, aluminum trim, white lettering; Chrome Green body with tangerine trim, white lettering; yellow body, black trim, red lettering; China Gold body, tangerine trim, ivory lettering; and ivory body, Chinese Gold trim, red lettering.

At the August 14, 1937, parade of Oliver farm equipment at the annual Oliver tractor plant employee picnic in Charles City, Iowa, "Six of the prettiest young ladies in the Oliver organization" were chosen to drive the half-dozen specially painted tractors in a parade of equipment that stretched for 2 miles.

Although the results were never made public, Oliver relied on the Chrome Green body, red trim, ivory letter combo into the 1950s.

A TIME OF CHANGE

In the 1950s, Oliver's engineers and management recognized the increasing trend toward larger horsepower tractors. They responded with plans for the 1800 and 1900 models to set the company's tractor criteria for the 1960s.

These tractors continued with the design innovations found in the Fleetline series. Plus they provided more horsepower, draft control, larger fuel tanks, and four-wheel-drive capability.

To join the line later would be the 1600 and 1750. These were smaller units based on the 1800 design. The 1800 used an upgraded engine from the 880 model, introduced in 1958, with larger displacement and higher rpm.

The 1900 would fall in the 100-PTO class and was designed to be the company's Wheatland tractor. The 1900 powerplant was a GM two-cycle four-cylinder diesel.

Design work on a suitable draft control system eventually led to a two-point hitch with lower link draft sensing for semimounted implements.

The design criteria for more fuel capacity were met by incorporating supplementary fuel tanks on the wheel guards—or "fenders" for us old-timers.

The tricycle-style 880 tractor was manufactured for only three years. This is a 1961 model. Engine speed was increased from 1,600 rpm on its predecessor, the Super 88, to 1,750 rpm on the Model 880 while the bore and stroke stayed the same. The Oliver-built six-cylinder engine produced 64 PTO horsepower.

123

OLIVER ENGINEERING

Concerning the state of engineering at Oliver in the 1960s, Vincent P. Weber, senior project engineer, says that, "During the 1960s engineering was located in Charles City, Iowa. Tractor engineering was located right there in the middle of the tractor plant. Engineering left Charles City in August 1970 and moved to Hopkins, Minnesota.

"The Charles City plant was a highly up-to-date facility in the 1960s too. We had a lot of numerically controlled machine tools, for example. We had big plans in the 1970s, but the economy spoiled those plans. We were looking forward to having a modern electric foundry, part of which was built although never fully utilized.

"We were working on higher horsepower engines, shift-on-the-go transmissions, and better hydraulics. Beginning in 1960, we came out with the larger tractors that were newer than the Fleetline. These were the 1800s, 1900s, and the 1600s. We'd started designing them in 1958 to 1959.

"In 1967, we got computers in engineering on time-share with IBM. One of the first things I got involved with as an engineer at Oliver was gear design. I designed the helical gears in the 880. That was back in the 1950s when I used a rotary calculator.

"It would usually require seven to eight hours of running the cutting data and tooth proportions for design of a pair of gears. Then, of course, the checker probably spent an equal amount of time checking the numbers before using them.

"Later we got electronic calculators. That halved the design time. Then after we got the computers and got a program set up, well, golly, we could obtain this data in only three minutes!

"We remained separate engineering departments at Oliver, Cockshutt, and Minneapolis-Moline until 1970. The tractor engineering was separate and it remained separate until it was moved to Hopkins, Minnesota."

This 1964 1600 model mounts a 1610 front-end loader and is still being used. Its engine puts out 60 horsepower. Horsepower was provided by the Oliver-built six-cylinder engine that displaced 231 cubic inches in the gasoline version or 256 cubic inches in the diesel model. The Model 1600 was built from 1962 through 1964.

The 1800 and 1900 were available in four-wheel drive via a mechanical front-wheel-assist design. Weber remembered, "The 1800 and 1900 had a Clark mechanical front-drive axle. I was involved in getting it fitted to the tractors. Later on we came up with a hydraulic drive that wasn't too successful nor popular. It cost too much and was power-consuming as well."

Fitted with Goodyear's Terra tires, these tractors provided low ground pressure compared to regular tires and so compacted soil less.

The regular six-speed transmission was supplemented by a Funk Reversomatic. A separate hand lever made shifting from forward to reverse possible without clutching.

In November 1959, the 1800 and 1900 tractors were introduced at the Hippodrome in Waterloo, Iowa. Initial sales were far better than anyone expected. The numbers on the production run had to be increased three times by mid-1961.

Yet almost by the time these tractors were home from the Waterloo show the company was under new ownership.

In 1960, White Motor Corporation bought some of Oliver's plants, including the plant that manufactured tillage equipment in South Bend, Indiana; the hay equipment plant in Shelbyville, Illinois; tractors in Charles City, Iowa; harvesting equipment in Battle Creek, Michigan; and seeding equipment in Springfield, Ohio. The name changed again, this time from the Oliver Corporation to Oliver Corporation, a subsidiary of White Motor Corporation. The remaining plants became a part of Amerada Hess.

Finally, in the early 1960s, White also bought the Cleveland line of crawlers (Cletrac) and the contents of the plant, which they moved to Charles City, Iowa. This, too, became a part of Oliver Corporation.

In 1962, White Motor Company purchased Cockshutt of Canada and made it part of Oliver Corporation. The following year, White bought

Minneapolis-Moline and operated it as a subsidiary separate from Oliver Corporation.

Although Oliver was owned by White during the 1960s, the Oliver name and badge continued to appear on a wide line of tractors. To fill the slots for horsepower ranges that the company didn't produce, it began to outsource from David Brown of England and Fiat of Italy.

Oliver's 500 was manufactured by David Brown in England and sold by Oliver in the United States with Oliver colors, decals, and grilles. It was basically David Brown's Model 850. The Oliver 500 was available with gasoline or diesel engine. It was introduced in 1960 and ran until 1964.

Also new in 1960, the 440 was a reincarnation of the Super 44 (produced in 1957 and 1958) and arrived with few, if any, changes from its predecessor. Sales numbers weren't enough to keep it in the line for long, and it's unclear when it was abandoned. Several sources list it as being produced only in 1960, yet there's some evidence that the 440 was also built in 1962.

The big news for Oliver in 1960 was the four-number tractors that brought new styling, plus a host of other features that made them as modern as any tractor a farmer could buy and put in his fields.

The 1800A diesel checked in at 61-drawbar horsepower from a six-cylinder Oliver engine. The engine was also available in gasoline or LP gas. Tricycle row-crop and adjustable wide-front axles were offered. The 1800A was upgraded in 1962 to the 1800B and again in 1963 to the 1800C.

Barney Retterath worked at Oliver in Experimental Testing and commented about the 1800 and 1900 series: "The C tractors were a more operator convenience-oriented tractor, with more features to make them easier to drive. For instance, there was the telescoping and tilting steering wheel."

The 1900A, also a 1960 offering, was available in Wheatland and Riceland models only. It used the GM four-cylinder, two-cycle diesel engine equipped with a blower that generated 80-drawbar horsepower. Both the 1800 and 1900 had six-speed transmissions. In 1962, it was designated as the 1900B with a Hydra-Power Drive. In 1963, it became the 1900C.

"I was involved in the Power-Shift section of the transmissions for the 1800 and 1900," Weber recalls. "We called it the Hydra-Power Drive. It provided

The Model 1650 tractor was versatile. It could pull heavy field equipment or it could be used for towing a long trailer picking up and delivering tomatoes to market. Produced from 1964 through 1969, the 1650 had the added versatility of Riceland, Row-Crop, Wheatland, or Hi-Crop models.

These three 1800s are set up to get plow-down fertilizer socked into the ground for the next corn crop in a hurry. All three are mount duals, not too common for the decade, with each outside dual running backward. Note what today would be regarded as primitive protection from the chill wind.

The 1900, like this 1960 Wheatland model, was respected as an impressive workhorse in its day. The downside was that it was so noisy that the driver would have to wear ear plugs or risk hearing damage.

shift-on-the-go. We had a unit in the Fleetline series, the 770 and 880, which we called the Power-Booster Drive. It wasn't too successful, though.

"The Hydraul-Shift was developed toward the end of the 1960s. It took the Hydra-Power Drive, which was a two-speed, and made it into a three-speed. With Hydra-Power you just down-shifted and then back to a direct drive.

"The three-speed added an overdrive. So we had three speeds for shift-on-the-go, allowing a choice of 18 speeds at a time. We had a basic six-speed transmission too. With the Hydra-Power, we had twelve speeds that could be shifted into without clutching.

"The Funk Reversomatic was built by the Funk company in Coffeyville, Kansas. It was used primarily on industrial tractors and was like a shuttle transmission on industrial vehicles.

"The Power-Booster would down-shift, like a low side. The main difference between that and the Hydra-Power was that it had a dry mechanical clutch.

"The Hydra-Power Drive became a self-contained unit with a wet hydraulic clutch. An overrunning clutch was used for the low range. You disengaged the direct-drive clutch and the overrunning clutch would then drive at a lower speed. Both the Power-Booster and the Hydra-Power Drive worked on this principle.

"The three-speed Hydraul-Shift added a planetary gear set and a second hydraulic clutch for overdrive."

Another David Brown–made Oliver tractor was the 600. In the line from 1962 to 1964, it was in the 43-drawbar horsepower class.

Just a bit more powerful was Oliver's 1600, made from 1962 to 1964. Available with either a diesel or gasoline six-cylinder engine, it developed 48-drawbar horsepower.

Made by Fiat in Italy, the Oliver 1250 was a 1965 offering that developed 38 horsepower from a four-cylinder Fiat-built diesel engine. The 1250 didn't meet expectations. It was judged unsatisfactory until Oliver engineers made some suggestions for redesigning the tractor.

Retterath remarked, "We did do some changing on the Fiat tractor, basically operator controls and PTO lever placement." The 1250 was in the Oliver line through 1969.

Another Fiat/Oliver was the Oliver 1450. It was introduced in 1967 and was offered through 1969. It, too, used the Fiat four-cylinder diesel engine that the manufacturer estimated at 55 horsepower.

Two other Fiat-built models were also available in the 1960s, the 1255 from 1969 to 1971 and the 1355 from 1969 to 1971.

New models in 1964 included the 1650, 1750, 1850, and 1950. With the exception of the Model 1950, which bowed out in 1967, these tractors continued in the line throughout 1969.

The diesel version used a 354-cubic inch six-cylinder Perkins engine. The gasoline version used the Waukesha Oliver engine. According to Weber, engineering hadn't yet turbocharged the Waukesha engine and so decided to go with the Perkins to get the horsepower the engineers felt farmers wanted.

"That GMC four-cylinder Detroit Diesel engine in the 1900 was a screamer and would roar like the dickens," said the owner of a 1960 Wheatland. Horsepower varied between 90 and 105 from 453 cubic inches, depending on where the pump was set.

Derived from the earlier 1800 series tractors, the 1750 was made in both gasoline and diesel versions. This 1965 tractor's diesel engine develops 87 horsepower from 310 cubic inches. It's also equipped with Hydra-Power Drive and has an adjustable wide front end.

127

The last 1850 made in 1969, this tractor is unusual in combining a six-speed transmission and over/under hydraulic shift with a Perkins diesel engine. Its horsepower was rated at 92.

The 1850 gas narrow front was a popular tractor; however, this 1967 model with six-speed over/under hydraulic shift the first year it was featured was rare since Oliver ceased marketing gasoline tractors about this time. The engine puts out 92 horsepower.

Concerning the 1850 engine, Retterath recalls, "The 1850 diesel was a Perkins and I actually put the first Perkins engine in an 1850. That engine had some trouble, though.

"The reason was that it originally used a pressed-steel valve cover. The pressed-steel covers would sometimes leak a great deal of oil. There was no more trouble after switching to a cast-iron cover."

Available with four-wheel drive, the Oliver 1950T came on board in 1967 and stayed through 1969. Equipped with a turbocharged Oliver six-cylinder engine, it produced 92-drawbar horsepower. Styling echoed previous models, but featured the auxiliary fender fuel tanks.

Oliver's Model 1550, introduced in 1965 and offered through 1969, was the exact same tractor as the later Minneapolis-Moline G550. Both diesel and gasoline versions were available in an Oliver-built six-cylinder engine.

The years 1968 and 1969 brought the 2050 and 2150 into Oliver's tractor fleet. The 2050 developed a maximum of 118-PTO horsepower from Oliver's six-cylinder diesel. The 2150 was the same tractor equipped with a turbocharger. This boosted the maximum PTO horsepower to 131.

By the time these tractors were introduced Oliver had perfected its over/under Hydraul-Shift feature to go with the standard six-speed transmission. Engaging the Hydraul-Shift into overdrive increased ground speed by 20 percent while decreasing pulling power by 17 percent. The underdrive side reversed the effective travel speed and pulling

The four-cylinder turbocharged GM diesel in the 1965 Oliver Model 1950 gives it 105-PTO horsepower. As its owner pointed out, "If you can stand the noise it's fun to drive. It runs cheaply, but does it make a lot of noise." Some of the 1950 Oliver models used Oliver's six-cylinder turbocharged engine that gave the same 105-PTO horsepower.

129

There weren't too many 2050s made, such as this 1969 model. The same 478-cubic inch Caterpillar V-8 engine that when turbocharged produced 135 horsepower in the larger 2150 developed 118 horsepower in the 2050.

power. In effect it gave a wide 18 forward speed range to the transmission.

The integrating of Oliver and Minneapolis-Moline models became apparent in the 55 series that was introduced in 1969. The Oliver 1555 and Minneapolis-Moline G550 were essentially the same tractor. The engine was a six-cylinder Oliver available in diesel, gasoline, or LP gas.

The Oliver 1655 offered the same three fuel options, plus three auxiliary transmission choices: Hydra-Power Drive, over/under Hydraul-Shift, or creeper drive. The 1655 was factory rated in the 70 horsepower class.

Sharing the same specs under different colors and badges was the Oliver 1855 and the Minneapolis-Moline G940.

Four row-crop versions of the Oliver 1955 were available, in addition to Wheatland, Riceland, and four-wheel-drive models. The turbocharged Oliver six-cylinder engine put the 1955 in the 108-PTO category. It could have either the diesel or gasoline engine.

"Factory-installed cabs were available during the 1960s and were purchased from several different vendors," Weber said.

Oliver, Cockshutt, and Minneapolis-Moline were merged into one entity, the White Farm Equipment Company, Division of White Motor Corporation, in late 1969. The company was headquartered in Hopkins, Minnesota.

The Charles City, Iowa, plant was closed in 1992 after 86 years of tractor manufacturing. An auction was held in 1993 to dispose of tooling and material, and in 1995 the plant was demolished.

MINNEAPOLIS-MOLINE

The Minneapolis-Moline Power Equipment Company, formed in 1929, was another example of three agricultural implement companies maneuvering to strengthen their position in a desperate industry that had just weathered some of its most trying years ever. Forming the new company were Moline Plow Company, Minneapolis Steel and Machinery Company, and Minneapolis Threshing Machine Company.

The Moline Plow Company began in Moline, Illinois, in 1852 as a manufacturer of hay rakes, pumps, and various other implements. It turned its

The top-of-the-line 2150 was introduced in 1968 and remained in the line only through 1969. A 2150 with front-wheel assist like this 1968 model was rarely seen. Horsepower from Oliver's turbocharged six-cylinder 478-cubic inch engine is 135 with 131 at the PTO.

A rather rare tractor is this 1969 diesel 1755, and it also has only 1,500 actual hours recorded on its tachometer. Horsepower was rated at 85.

131

"Quiet and fun to drive" is how the owner of this 1969 1655 describes what he calls his favorite driving tractor. The 1655 was made from 1969 to 1975. Its gasoline engine was rated at 66 horsepower. Oliver's famous six-cylinder engine was available in either gasoline or diesel for the Model 1655.

Reported to be a "nice driving tractor," this 1969 Oliver Model 1955 has worked only 1,800 actual hours. It's rated at 105 drawbar horsepower and 108 PTO. Various Model 1955s were produced from 1967 through 1974. They all used the Oliver turbocharged six-cylinder engine available only in the diesel version.

attention to plows in 1865 and became a well-known maker of plows for the Plains states.

To make it a truly full-line company a motor plow was designed, and International Harvester Company was contracted to build five prototype units. This effort was soon abandoned, so in 1915 Moline Plow purchased the Universal Tractor Company of Columbus, Ohio. The initial offering was a two-cylinder tractor. Then in 1918 the four-cylinder Moline Universal debuted. It was marketed until 1923.

The Universal was a departure from the standard rear-wheel-drive design. The Universal featured two large drive wheels located in front with the engine over the front axle. In concept and design it was similar to the Allis-Chalmers 6-12 of the same time period.

Neither tractor offering really appealed to farmers. And, besides, a Fordson tractor could be bought for a lot less money. The improved Universal D came out in 1917. One year later it featured, as standard equipment, an electric self-starter and an electric headlight.

Financial problems plagued Moline Plow in the 1920s. It incorporated as Moline Plow Company Inc. in 1922 and reorganized by 1923 as Moline Implement Company, selling off a number of assets in the process. These drastic measures allowed the firm to survive and become a viable agricultural implement manufacturer until it became part of Minneapolis-Moline.

The second player in the story, Minneapolis Steel and Machinery Company, was organized in 1902 as a fabricator of steel components for large structures such as bridges and water towers. Shortly, the company began to explore the stationary steam engine market and then switched to gasoline-powered units.

From 1906 to 1910 the company promoted its gasoline engines to firms such as J. I. Case Threshing Machine and Reeves and Company, Columbus, Indiana. The latter company used the engines in its own 40-horsepower tractor.

In 1910, Minneapolis Steel took the plunge into tractor manufacturing by contracting with the Joy-Wilson Company, Minneapolis, Minnesota, to design and build a tractor for the company. After the initial five prototypes, Joy-Wilson produced the Twin City Model 40 tractor for Minneapolis Steel through 1920.

Minneapolis Steel contracted with J. I. Case Threshing Machine to build 500 Case 30-60 tractors during 1912. While Joy-Wilson was manufacturing the Twin City for Minneapolis Steel, the Bull Tractor Company of Minneapolis contracted with Minneapolis Steel to build 4,600 Little Bull and Big Bull tractors. This rather strange manufacturing arrangement apparently lasted until 1916, when Minneapolis Steel withdrew from production of the

The engine of the 1960 4 Star Super provided 48-PTO and 44-drawbar horsepower. The tractor included all the many little improvements that were obtained by tweaking and fine-tuning the previous 445 model.

133

JEEP'S TIES TO MINNEAPOLIS-MOLINE

John N. Willys was a large stockholder in the Moline Plow Company by 1914, but the controlling stock block was owned by the Stephens family that also owned the Moline Automobile Company. The latter produced the Moline-Knight automobile that year. So, in effect, Moline Plow was in the automobile business since one family controlled both companies.

In 1938, Minneapolis-Moline engineers used the engine from the Comfortractor to power the UTX, an all-wheel-drive vehicle designed for military use. Tested at Camp Ripley, Minnesota, the UTX acquired the nickname Jeep.

Yet when all the government bidding and politicking was done, Willys-Overland Motors and Ford Motor Company got the government contracts for the military "go-anywhere reconnaissance vehicle."

For whatever it's worth, though, the government did give Minneapolis-Moline credit for the name Jeep. When others, especially Willys, began using the name in promotional material, Minneapolis-Moline hollered foul.

The House Committee on Military Affairs then got involved and in 1942 validated Minneapolis-Moline's claim to the name. Of course Willys, perhaps in a way, assumed he had some right to use the name since he was a major stockholder in Moline Plow, which became part of Minneapolis-Moline.

All engines in Minneapolis-Moline tractors up until this point had a maximum of 283 cubic inches. The 1961 M5 broke new ground with 336 cubic inches, thanks to its 4 5/8x5-inch bore and stroke engine. This greater capacity gave it 57-drawbar and 61-PTO horsepower.

Bull tractors. Minneapolis Steel did profit from the experience of building the smaller Bull tractors. In 1918, it brought out its own smaller Twin City 16-30. A year later the 12-20 Twin City appeared with a futuristic four-cylinder engine featuring dual camshafts and four valves per cylinder.

The Minneapolis Threshing Machine, the third component in the Minneapolis-Moline story, moved to Minneapolis from Fond du Lac, Wisconsin, in 1887. Beginning in 1910, Minneapolis Threshing began offering the Universal 20-40 tractor in its product line. This two-cylinder gasoline model was made by the Universal Tractor Company of Stillwater, Minnesota.

Before long, Minneapolis Threshing decided to manufacture its own tractors. In 1911, it offered its own 25-50 four-cylinder model. The following year it brought out the 40-80 (later rerated as 35-70). These tractors carried the name "The Minneapolis," "Minneapolis," "MTMC," or some combination.

In 1914, the 20-40, and in 1915, the 15-30, were introduced to its tractor line. Responding to the demand for smaller tractors, Minneapolis Threshing also introduced a smaller line in the 1920s.

A hallmark of these tractors was the long-stroke engine that provided great lugging power when heavy loads were encountered in the field and on the belt. This feature was carried over into the Minneapolis-Moline line.

Since each of the three companies that formed Minneapolis-Moline Power Equipment Company already had tractors in their product lines, the new company was forced to choose which tractor models to scrap and which to continue marketing.

It chose to offer the Minneapolis line until inventories of the 17-28 and 27-44 were sold. Both models were gone by 1935.

The Twin City line stayed, and during 1934 the KT, MT, and FT were replaced with the improved KTA, MTA, and FTA versions. Also in 1934, the Universal JT row-crop joined the line and stayed through 1937. Additional offerings included the Universal Z in 1936 and the Universal ZN in 1938. Variations were the JTO Orchard and a standard model of the Z.

"Visionline" design was featured with the introduction of the Universal Z. The concept was to streamline the appearance of the tractor while at the

A 1964 Minneapolis-Moline Model M670 Super equipped with a TW900 two-way spinner plow. Drawbar horsepower was 64, with PTO horsepower at 73. The first year for the Model M670 was 1964, with production continuing through 1970. The four-cylinder Minneapolis-Moline engine was available in gasoline, diesel, or LP gas.

Powered by LP fuel, this 1964 M602 gave 54-drawbar horsepower when hitched to an implement and 64-PTO horsepower. Available for model years 1963 and 1964, the M602 could also be equipped with a gasoline or diesel version of the Minneapolis-Moline–built engine.

same time improving operator vision. This step in approving the appearance and salability of the line included a change in color scheme. The company selected Prairie Gold as the color for the new Minneapolis-Moline tractors.

In 1938, Minneapolis-Moline unveiled its revolutionary new UDLX Comfortractor and Sport Open Model Roadster. This latter model was a Comfortractor minus the cab. These tractors actually encroached on automobile design in an effort to combine the functions of both into one multipurpose vehicle. Fitted with an integral, factory production cab, the UDLX Comfortractor sported windshield wipers, heater, radio, roll-up windows, and seating for three. Fifth gear was capable of 40 miles per hour on the road. After a day in the field the farmer could speed to town for shopping or entertainment! It was an exciting concept. Unfortunately, it didn't sell and only approximately 150 were ever produced. The cab was an option available on the Minneapolis-Moline Model R row-crop introduced in 1938.

Another model that received the Visionline styling was the Model U in 1938. The U fell into the

Coming out in 1964, this four-wheel-drive G706 was one of the first Minneapolis-Moline tractors to provide three-point hitch capability. Rather than internal regulation, two auxiliary hydraulic cylinder were used for positive action both up and down. The 504-cubic inch diesel engine was rated at 89 horsepower at the drawbar and 110 horsepower at the PTO.

four-plow class with five forward gears, disc brakes, and oil-bath air cleaner. Customers could choose between either a low-compression or high-compression four-cylinder engine. The U series tractors were offered for 20 years, becoming Minneapolis-Moline's best-seller at more than 85,000 units.

The big horsepower tractor for Minneapolis-Moline in 1940 was the G, which, with some upgrades, stayed until the mid-1950s. The GT and GTA ran until 1947, followed by the GTB from 1947 to 1951. The GT tested right at 50-drawbar horsepower. The GTC was the LP gas version of the GTB. Beginning in 1941, Minneapolis-Moline was one of the first, if not the first, to offer LP gas as a fuel option on tractors.

In 1951, Minneapolis-Moline purchased B. F. Avery to add production capacity and to extend its marketing network into the southern states, where the Avery name had been around for more than 125 years. Avery started as a plow manufacturer but added small tractors designed primarily for the tobacco farmer. Avery's Model V was a single-plow tractor and its BF a two-plow all-purpose unit that Minneapolis-Moline kept in its line. The Avery name was added to the company's tractors for a short while and later dropped.

Minneapolis-Moline's Powerline series was introduced in 1953 with the 335, 445, and 5-Star models. These models illustrated a major upgrade in styling and features. New power steering, transmissions, and hydraulics were offered that made these tractors state-of-the-art when introduced. Horsepower ratings showed 30 for the 335, 40 on the 445, and 50 for the 5-Star.

Made in 1965, this U302 came with a 220-cubic inch engine. It put out 49-drawbar and 55-PTO horsepower. This tractor was an early provider of a combination of operator rollover protection and sunshade.

136

MINNEAPOLIS-MOLINE IN THE 1960S

The 1960s didn't open well for Minneapolis-Moline. Annual net sales for the company had dropped from a 1953 record high of $107 million to $49 million in 1960. The boom years of the 1950s never returned for the company.

Lee Vaughan retired as tractor project engineer at Minneapolis-Moline after 40 years in the farm equipment industry. "During the 1970s, Minneapolis-Moline's engineering was located at Hopkins, Minnesota," Vaughn explained. "The design work was done in Hopkins and the tractors were made at the Lake Street plant on the south side of Minneapolis.

"The engine and powertrain test facility was also located at Lake Street; however, there really wasn't any acreage available for field testing at the Lake Street facility. So field testing was actually performed off site. Some was done in Phoenix, Arizona, but the majority was done by placing the tractors with large growers like Green Giant in Wisconsin and Minnesota."

Commenting on engines, Vaughan says, "We did pretty much of our own engine manufacturing. We also had a foundry located at Lake Street where we could do our own castings. The engines were a

Manufactured in 1966, this Jet Star III put out 41-drawbar and 44-PTO horsepower from a 206-cubic inch engine.

The 1969 G1050 carried a six-cylinder engine of 504 cubic inches that was powered by propane. Horsepower was 101 at the drawbar and 111 at the PTO. A diesel engine was also available.

Proudly displaying red, white, and blue, the 1969 G950 diesel was a clear break with Minneapolis-Moline's traditional colors. It was also known as the Heritage Model and only a limited number were produced. Although made at a Minneapolis-Moline plant and carrying the M-M name, the company was under the ownership of White by this time. It had 86-drawbar horsepower and 97-PTO horsepower.

A GIANT INNOVATIVE STEP IN THE WRONG DIRECTION

Martin Ronning, chief engineer of Minneapolis-Moline, conceived the idea in the 1940s, but it wasn't until 1955 that the end result reached the marketplace. The concept, and subsequent product, earned Ronning the American Society of Agricultural Engineer's gold medal in 1956.

Ronning's idea materialized as the Minneapolis-Moline Uni-Tractor, a basic chassis and power unit capable of being readily transformed into any of several self-propelled agricultural implements.

Original attachments were the Uni-Foragor (forage harvester), Uni-Harvestor (combine), and Uni-Picker (corn picker). Later the Uni-Baler (hay baler) and other attachments were brought on line.

There's some indication that company market research didn't give the Uni-Tractor high marks. Later sales figures validated the research. The Uni-Tractor and its line of attachments was offered until 1962, when it was bought by New Idea of Coldwater, Ohio.

Harry Ferguson also tried this same concept with similar results when farmer acceptance just didn't materialize. Hindsight is always perfect. No doubt Minneapolis-Moline would have benefited much more if the considerable amount of engineering time and company dollars would have instead been spent on developing and refining its tractor line for the 1960s.

totally in-house operation, including design, casting, machining, and testing.

"Oliver had used some Perkins engines. So I tried out a Perkins one time; however, we stayed with the four-cylinder engine in the 1950s and the six-cylinder that I was working with, which had the three individual heads.

"Transmissions were a big issue in the 1960s, and we were getting into making our own transmissions. The company was working on a Power-Shift transmission and spent a lot of money on that."

Was Minneapolis-Moline engineering affected when White purchased the company? Vaughan believed that it was, noting that "both Minneapolis-Moline and Oliver were quite innovative in the 1940s and 1950s. They brought out features such as LP gas tractors, electric hydraulic controls, and a high-compression engine.

"When White Motor took over, its management was trying to develop its own engine design. White wanted to put a common engine in all the

The 1969 Heritage model A4T-1600 was powered by a 585-cubic inch diesel engine and sported red, white, and black rather than the usual yellow or Prairie Gold paint scheme. Its engine produced 129 horsepower at the drawbar and 143 horsepower at the PTO.

Minneapolis-Moline, Oliver, and Cockshutt tractors. White had an engineering facility for engines in California that was working on development and advanced engine design."

In a way, White engineering was on the move. Vaughan recalled, "In the late 1960s I went down to Charles City to work with the Oliver people. In 1970, they moved all engineering to Hopkins, Minnesota. Then again, in 1975, they moved all engineering to Libertyville, Illinois, right north of Chicago."

In 1960, the M-5 series was brought on-line to replace the 5-Star tractors. The series was available with gasoline, diesel or LP gas engines. The M-5 equipped with a front-wheel-assist axle was designated the M-504. The M-5 Diesel produced 50-drawbar horsepower from a four-cylinder Minneapolis-Moline engine.

For 1962, a Minneapolis-Moline six-cylinder diesel engine placed the G705 in the 90-drawbar horsepower range. The G705 was also available in LP gas. Its transmission provided five forward gears. The G705 model lasted through 1965.

Sharing the same engine and features as the G705, the G706 offered power-assist front wheels. It was introduced in 1963 and continued through 1967. The G706 was the exact same tractor as the Massey-Ferguson MF97 that Minneapolis-Moline manufactured. It had Massey-Ferguson colors and nameplate.

The G707 was an upgraded version of the G705. Both were two-wheel-drive–only vehicles. The G708, equipped with power-assist front wheels, was an upgraded version of the G706.

Introduced in 1964 and falling in the 48-drawbar horsepower area, the U-302 was powered by either a four-cylinder gasoline or LP gas Minneapolis-Moline engine. It had a transmission with 10 forward speeds and was offered through 1972.

The M-670 was available in a gasoline or LP gas, four-cylinder Minneapolis-Moline engine that was rated at 65-drawbar horsepower. It replaced the Model 602 series. The changes were hydraulic features and sheet metal. A 1964 through 1970 offering, it used the 10-speed transmission. It was also available as a Super M-670 model.

Equipped with either a diesel or LP gas engine, the G1000 replaced the 705 and 706 models for 1965 to 1969. This was a 100-horsepower tractor using a six-cylinder Minneapolis-Moline engine in either diesel or LP gas versions.

This 1969 Minneapolis-Moline A4T-1600 Plainsman Heritage model developed a hefty 128-drawbar and 143-PTO horsepower from its huge 504-cubic inch propane-powered engine. This particular tractor was the 40th from the last one made.

This 1969 G1000 easily pulls three tandem disks hooked together with a special hitch thanks to the 102 horsepower its 504-cubic inch engine put out at the drawbar. Although installed primarily to reduce compaction, the exceptionally large tires also provided excellent traction.

Still in its original paint and carrying a factory-installed cab, the G1000 Vista was a refinement of the G1000 Series tractors in 1969. The seat was raised on the platform and "provided terrific visibility," according to its owner. Operating a 504-cubic inch engine, it produced 102-drawbar and 111-PTO horsepower.

There are tires, and then there are tires! The special wheels and huge tires were an attempt to lessen the effects of the weight of a heavy tractor and rather narrow tires that cause soil compaction. Crop plant roots find compaction difficult to penetrate, and so yields usually suffer.

The G1000 Vista replaced the G1000 in 1969 as a diesel-only model. The Vista positioned the fuel tank behind the operator's platform, which was mounted on rubber pads.

Minneapolis-Moline's first row-crop tractor to approach the 100-horsepower mark was the G950, introduced in 1967. It was powered by a six-cylinder Minneapolis-Moline engine with the buyer's choice of gasoline, diesel, or LP gas fuel. The LP gas model carried the G900 designation. The gasoline and diesel models incorporated fender fuel tanks that increased the fuel capacity to 118 gallons.

In 1969, the G1350 was introduced. It tested out at 125-drawbar horsepower from a 585-cubic inch, diesel, six-cylinder, Minneapolis-Moline powerplant. Minneapolis-Moline produced this tractor for Oliver as its Model 2155 diesel. LP gas power was an option.

The A4T-1600 was an articulated four-wheel-drive tractor providing 127-drawbar horsepower. The transmission had 10 forward speeds behind a six-cylinder Minneapolis-Moline engine. This tractor was also marketed as the Oliver 2655 Diesel, with Oliver colors and badge, and as the White Plainsman. The

A4T was manufactured at White Equipment Company's facility in Minneapolis.

In 1963, the White Motor Corporation purchased a controlling interest in the Minneapolis-Moline Company; however, the badge and colors of the famous Minnie-Mo tractors would continue throughout the 1960s.

COCKSHUTT

The Brantford Plow Works, Brantford, Ontario, Canada, opened for business in 1877 as a blacksmith manufacturer of plows, cultivators, rollers, and planters. James G. Cockshutt was the proprietor and driving force behind the venture.

He incorporated the business as the Cockshutt Plow Company in 1882, but he died of tuberculosis only three years later. The company endured, going public in 1910. The company expanded, acquiring the Adams Wagon Company, Brantford Carriage Company, and Frost and Wood. These additional companies expanded Cockshutt's equipment line to include binders, mowers, rakes, carriages, and wagons.

The company marketed Hart-Parr tractors under the name Cockshutt Hart-Parr from 1924 to 1928, when it replaced them with Allis-Chalmers models. In 1934, the company switched back to Hart-Parr, which had become part of Oliver.

Warren Wheeler is a tractor collector and serves as a director of the International Cockshutt Club. Wheeler explained how this arrangement worked, pointing out that "just about the entire line that Oliver produced was sold by Cockshutt in Canada. Apparently Cockshutt had an exclusive arrangement with Oliver to sell tractors in Canada, because there were no Oliver dealers in Canada during this time.

"This arrangement continued through the Oliver 60, 70, and 80 models that were painted red and marketed in Canada as the Cockshutt 60, 70, and 80. In all those models there were certain cast parts that said 'Cockshutt.' They were sold in pretty big numbers, but strictly in Canada.

"The arrangement ended before the Fleetline Olivers were introduced. At this time Cockshutt went in-house with its tractor production."

Made in 1960, this Cockshutt Model 540 houses a 38-horsepower, four-cylinder Continental gasoline engine. The 500 Series tractors were introduced in 1958 and were a product of Cockshutt's plant at Brantford, Ontario, Canada.

Showing the narrow front-end option, this 1960 Cockshutt Model 560 is powered by a Perkins four-cylinder 270-cubic inch diesel engine. No gasoline engine was available. Rated at 50-PTO horsepower, the Model 560 had a six-speed forward and two-speed reverse transmission. This tractor was in the line from 1958 to 1961.

Engineers began compiling a list of specs and features for the Cockshutt Model 30 in 1944 for introduction in 1946. Full production began in 1947 and ended in 1956. The 30 was Cockshutt's biggest seller with more than 37,000 tractors produced.

The next tractor, the Model 40, used a 230-cubic inch Buda four-cylinder engine available in gasoline, distillate, or diesel versions from 1949 to 1957. The 40 was sold in the United States as the Golden Eagle.

The first Cockshutt 20s used a 124-cubic inch Continental engine that was soon replaced by the larger Continental 140-cubic inch version. Both gasoline and distillate fuel engines were available, plus the option of live PTO and hydraulics. The National Machinery Cooperative, which Cockshutt purchased in 1952, sold this tractor as its Model E2. Production ran from 1952 to 1958.

For 1956 and 1957, the Model 30 tractors with the Buda engine were replaced with the Cockshutt Model 35, which featured a 198-cubic inch Hercules powerplant.

The agricultural market slumped in the 1950s, which placed Cockshutt in a weak financial position. In 1957, the English Transcontinental Company purchased a controlling interest in the company and changed its name to the Cockshutt Farm Equipment Company of Canada Ltd.

In 1958, Cockshutt restyled and redesigned its entire line with the addition of significant engineering upgrades. The 500 series tractors were introduced at this time with a variety of powerplants.

A six-speed transmission was standard for all models, as well as optional hydraulics and live PTO. The tractors were styled by industrial designer Raymond Loewy.

The horsepower heavy-hitter for Cockshutt was the D-50. It used a six-cylinder Buda engine that generated 47-drawbar horsepower. The options were diesel only, hydraulics, and live PTO. This tractor was seen in the Co-Op line as the E5.

THE 1960S FOR COCKSHUTT

The 500 series tractors made the decade transition and were in production at Brantford until White bought the company in 1962.

"All 1950s tractors were built in Brantford, Ontario, Canada. The company continued to build tractors at Brantford until White purchased Cockshutt in 1962," Wheeler said.

From then on Cockshutt tractors were really Oliver tractors with Cockshutt badges and colors.

The 540 Wide Adjustable had a Continental four-cylinder gasoline engine, while the 550 Standard Diesel used the Hercules four-cylinder that developed 40-PTO horsepower. Both tractors had a six-forward and two-reverse gearbox and were offered from 1958 to 1962.

The gasoline version, the 550 Standard, had the same tenure in the line and also used the Hercules four-cylinder with the same horsepower rating and other specs.

"Cockshutt was always just a little bit ahead of time with its designs," observes the owner of this 1960 550. In addition to being a three-bottom plow tractor with 48 gasoline engine horsepower, it also featured live PTO, a three-point hitch, and both adjustable wide front and rear axles. Both gasoline and diesel engines provided 198 cubic inches.

The 570 was the biggest tractor Cockshutt made from 1958 to 1962. Available with either gasoline or diesel engines, this particular unit carries a Hercules six-cylinder diesel engine of 298 cubic inches. It also has an adjustable wide front end.

Called a standard front-end model with nonadjustable front and rear axles, this 1960 570 Wheatland used a 298-cubic inch diesel engine. The rationale behind nonadjustable axles was that because it was built to primarily work in wheat fields, adjustments weren't a necessity. The big, wide fenders were supposed to keep dust from making the driver uncomfortable.

Considered an intermediate-sized tractor, the 550 was designed for chores and light fieldwork. This 1961 model uses a Hercules four-cylinder gasoline engine putting out 45 horsepower.

Rated in the 50-PTO horsepower class from a four-cylinder Perkins engine, the 560 Standard Diesel was introduced in 1958 and stayed in the line until 1961. It came with the six/two transmission.

A Hercules six-cylinder 298-cubic inch engine put the 570 Standard Diesel at 64-PTO horsepower. This Cockshutt tractor also came with the six/two gearbox.

Cockshutt's 570 Standard Gasoline model used the same engine and same gearbox as the 570 Diesel, but tested 4 fewer horsepower than the diesel. Both models were offered from 1958 to 1960.

During 1961 and 1962, the 570 Super Diesel featured the Hercules six-cylinder with 339-cubic inch displacement that upped the horsepower to 65 at the PTO. The transmission was also the six/two.

A second "go-round" Oliver tractor with Cockshutt badge and colors was the 770. It came on-line in 1958 and continued to be offered until 1967. It used the same six-cylinder engine as the Oliver Model 77 and was offered in gasoline, diesel, and LP gas.

In 1963, the Oliver 880 joined the lineup as the Cockshutt 880. Both the 770 and 880 came with Power-Booster drive that effectively gave the tractor 12 forward and 4 reverse speeds. Front axles included the narrow dual, single, or wide utility. The 880 was also offered as a Wheatland or orchard model.

According to Wheeler, "In 1966 and 1967, there was one model marketed as Cockshutt that was produced by Minneapolis-Moline in Hopkins, Minnesota. It was the Minneapolis-Moline Jet Star III Super in red-and-white Cockshutt colors and designated as the Model 1350. It does have some variants, although they're minor.

"It has the traditional Cockshutt spear down the side that you also see on Oliver tractors. The spear decal on the hood is a carryover from the Cockshutt line. Cockshutt used it starting with the 500 series tractors in 1958."

A four-cylinder Hercules gasoline engine let this 1961 Cockshutt Model 550 work with 48 horsepower (40 PTO). For those that liked a diesel powerplant, the Model 550 was available with the same four-cylinder Hercules engine in a diesel version.

After the Brantford plant was closed, the Cockshutt tractor line became the same as the Olivers. Of course, the paint schemes and badges would be appropriate for Cockshutt. So specs and features of the Cockshutt line paralleled those of the Oliver models described earlier.

Wheeler talked about Cockshutt's claim that it was the first company with live PTO: "Ivan McRae, chief engineer, tractor division at Cockshutt, had the patent for the live PTO.

"As far as marketing an effective live PTO as we know it today, mounted where it's mounted and used like it's used, with a continuous action with a clutch, that came from Cockshutt and was patented by Cockshutt."

THREE HISTORIC NAMES FADE AWAY

Gradually all individual company names, badges, and colors disappeared. After 1975, the names *Oliver*, *Minneapolis-Moline*, and *Cockshutt* no longer appeared on the tractors. The colors were changed to the two-tone gray of White.

White Motor Company itself fell upon hard times in the late 1970s. In December 1980, Texas Investment Corporation purchased the White Farm Equipment Company Division of White Motor Company. It was then operated as a wholly owned subsidiary.

Certain assets of White Farm Equipment Company were bought by Allied Products Corporation in November 1985. The following year White Farm Equipment purchased some assets of White Farm Manufacturing of Canada.

The next development in this ongoing buy-and-sell occurred in 1987 when White Farm Equipment merged with another Allied subsidiary to become White-New Idea Farm Equipment Company, with headquarters in Coldwater, Ohio.

Ownership changed again in June 1991, when AGCO purchased the White-New Idea Farm Equipment Company.

Oliver built this extremely rare red 1968 1650 Wheatland-style tractor for the Canadian market. It has a 66-horsepower engine; however, targeted at the prairie wheat-producing areas, it lacked a three-point hitch.

145

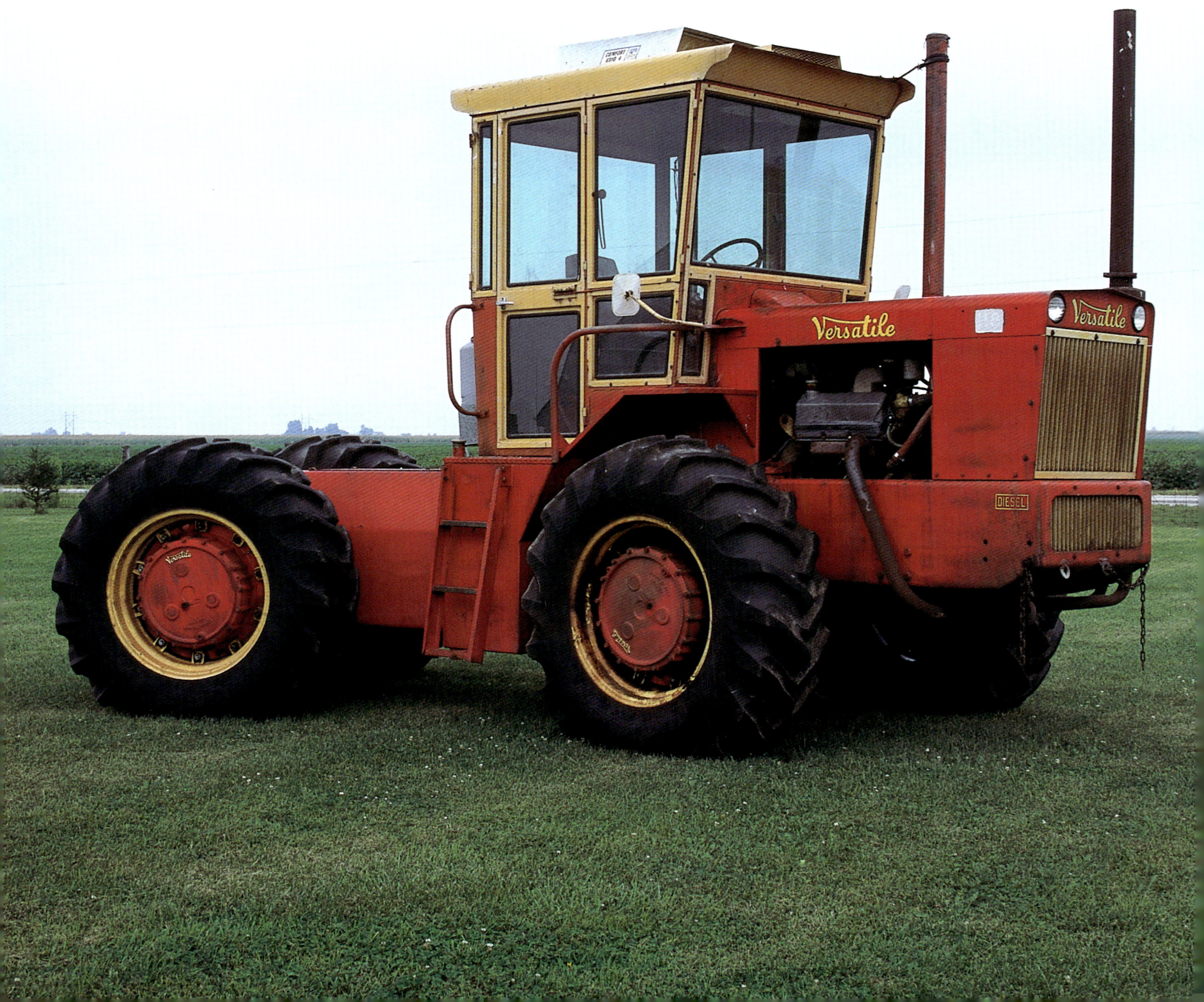
Versatile
Versatile
DIESEL

CHAPTER EIGHT

OTHER FARM TRACTOR COMPANIES

A study of the tractor industry up to the 1960s would convince most people that the world of tractor manufacturing wasn't a friendly environment.

The historical landscape was littered with so many manufacturers that had tried and failed. Fierce competition, economic cycles, labor woes, and the weather all seemed to conspire against the brave few companies that dared to challenge the odds.

Sometimes fate smiles on those who don't know something can't be done, however. In the 1950s and 1960s, several individuals didn't realize they couldn't succeed at launching a tractor manufacturing company. Ignoring the odds, they provided good products for niche markets.

The mega-horsepower, four-wheel-drive farm tractor established these firms as "players" in the farm equipment industry.

STEIGER

Douglas and Maurice Steiger just wanted a big tractor, bigger than any being offered by the industry in 1957. With the "solve your own problems" attitude typical of farmers, they rounded up a bunch of Euclid truck parts and a 238-horsepower diesel engine to build the first Steiger tractor in their farm shop at Red Lake Falls, Minnesota.

Their neighbors were greatly impressed with it. Before long the Steigers were fabricating copies of their tractor for the local market, which soon expanded to a national market.

The first Steiger tractor was powered by a six-cylinder Detroit diesel rated at 238 engine horsepower coupled to a five-speed transmission. The Steiger 1 FWD was made in both 1957 and 1958.

Their second generation was introduced in 1963 and included four models, the 1200, 1700, 2200, and 3300.

Company literature listed these features for the new series:
- New swinging power divider makes possible shorter turns while putting less strain on the driving joints
- New V-type diesel engines
- Live hydraulic systems
- All-gear drive
- Synchronized transmission
- Two-axle oscillation
- Balanced for drawbar work
- Available with cab
- Standard components

Featuring "standard components" reflected the company's policy of outsourcing major components such as engines, transmissions, and powertrains. Thanks to using major brand components, replacement parts were readily available.

The 1700 and 2200 used Spicer, Clark, and Steiger components in their powertrains. Main frames were one-piece welded construction of heavy-gauge steel.

The 1200 used a four-cylinder Detroit diesel engine and a 12-speed transmission.

When this 1967 118 was made the year after Versatile began producing tractors in 1966, its 145 horsepower was thought to make it a "big tractor." Power comes from a Cummins V-6 engine.

147

When the Steiger brothers first started producing a limited number of tractors in 1963 in a barn, they built them extremely heavy. For example, instead of the usual decals, the Steiger nameplate is cut from 3/8-inch steel. The big V at front indicates this 1967 2300 carries a V-671 Detroit Diesel 230-horsepower engine. It has a 10-speed transmission. Of the 16 tractors built in this series, only two left the factory with a cab.

Fitted with a V-6 Detroit diesel engine, the 1700 carried an advertised horsepower rating of 216. This model featured a nine-speed transmission, and the tractor weighed in at 19,000 pounds.

Steiger's 2200 used a bigger version of Detroit's V-6 diesel that boosted the advertised horsepower to 265. It also came with the nine-speed transmission and weighed 3 tons more than the 1700.

At the top of the horsepower class for the 1960s was the 3300 with a 9.3-liter Detroit V-8 diesel engine that was rated at 318 horsepower. The full range Power-Shift transmission had 16 forward speeds. This tractor tipped the scales at a hefty 31,000 pounds.

Marketed mostly in Canada, Montana, and the Dakotas, total sales amounted to approximately 120 units of these models until 1969. At that time the Steiger brothers, along with other investors, formed a new corporation and moved manufacturing to Fargo, North Dakota.

The first tractor built at the new facility in Fargo was called the Wildcat. Steiger chose the Caterpillar 3145 V-8 diesel that generated 175 horsepower as the engine for its first Wildcat.

Cummins engines also found their way into the Steiger tractors for the 800 series Tiger Diesel. The V-903 Cummins placed the Tiger in the 320 horsepower class. A 10-speed transmission was standard.

Steiger adapted some of its models for the logging industry too. The 850 and 1250 could be outfitted with dozer blade, canopy, and a "hot shift" reverse. A two-speed winch married to a torque converter provided winch speeds from 0 feet a minute to 350 feet a minute.

Starting in 1973 the company began to expand its foreign market operations. Steiger, Australia Ltd. was established in 1976 and was followed by dealerships in Central and South America.

This powerful Steiger four-wheel-drive tractor with quad duals is typical of the company's product during the late 1960s. The tractor has the horses to pull multiple pieces of equipment at the same time over large acreages. Like other such size tractors, though, it wasn't ideal for farming small tracts or fields with many terraces close together.

International Harvester began sourcing its four-wheel-drive tractors from Steiger in 1973. Then, in 1986, the Steiger company was folded into the Case International Tenneco organization.

VERSATILE

Tool designer Peter Pakosh, worked for Massey-Harris when, in his spare time, he built his first grain auger in his Toronto, Ontario, home's basement. His first 10 augers sold quickly. He built 50 units in 1946. A year later, Pakosh and his brother-in-law started the Hydraulic Engineering Company.

In 1951, they moved the company to Winnipeg, Manitoba, and began designing and manufacturing self-propelled swathers. Versatile Manufacturing Ltd. was incorporated in 1963.

Versatile built its first four-wheel-drive tractor in 1966. Designated the D100, it was powered by a Ford six-cylinder diesel engine and carried a manufacturer's estimated horsepower rating of 125.

This model was also available as the G100 in a gasoline version. Its gasoline engine was a Chrysler V-8 that produced 100 engine horsepower. These first units used four-speed gearboxes.

The D118 used a 5.7-liter Cummins V-6 engine classed at 140 horsepower. It was of 1967 vintage. With modifications and upgrades, the D118 was in production through 1971. The D118 transmission was upgraded to nine forward speeds.

The gasoline version of the 125 4WD introduced in 1967 featured a Ford V-8 that delivered 125-drawbar horsepower. Transmission range was nine forward and three reverse speeds. Production lasted through the 1970 model year.

Outfitted with a Cummins V-8 diesel engine, the 145 4WD was capable of 145-drawbar horsepower. It used the 9/3 transmission and was in production from 1967 through 1971.

Versatile also made pull-type combines in the late 1980s and early 1990s. Plus the company offered windrowers-swathers at the same time it introduced its tractor line.

Ford needed a high horsepower four-wheel-drive tractor for its line, and so bought Versatile in 1989.

CATERPILLAR

The words "Caterpillar" and "construction" have become virtually synonymous. From city street

The V emblem stands for Versatile or V-8 power, your choice; but it could also denote victory over just about any job that required hitching up to one or multiple pieces of large equipment. Like most later four-wheel-drive tractors, it sported eight wheels in an attempt to reduce soil compaction from its great weight.

Although not too plentiful in the 1960s, there were some track-type tractors used on farms. Caterpillar led in numbers of these. Incidentally, in just the past few years there's been a surge in interest in track-type tractors as soil compaction by heavy tractors becomes a yield limiting factor. While this D2 might look rather small if put alongside a wheel tractor also capable of pulling four plow bottoms, it supplied both the power and traction needed.

This Caterpillar D2 plows down sweet clover used as a nitrogen-producing rotation crop in the days when there was less reliance on commercial fertilizers. In common with all track-type tractors, there was no steering wheel.

Yes, a D6 "Cat" can have a life away from an industrial site too. This D6 was perfect for packing silage in a bunker silo: bulldozer blade, weight, ease of steering, and stability, thanks to its lower center of gravity.

improvement to mega projects such as the Hoover dam, the familiar yellow crawlers can be seen just about anywhere on earth.

Yet their genesis was in the vast agricultural fields of America's far West more than a century ago. Agricultural application continued through the decades but was always a small part of the company's focus, until recently, when Caterpillar is again associated with the agricultural industry.

Daniel Best's Best Manufacturing Company of San Leandro, California, was founded in 1885 to build steam traction engines and combines. Best put a gasoline tractor on the market in 1895. He was also the father of C. L. Best, a man who later played a significant role in the Caterpillar story.

The "other" firm with similar products and ideas was the Holt Manufacturing Company of Stockton, California, founded by Benjamin and Charles Holt. Holt tested its first track-type vehicle in 1904. It sold the first steam-powered crawler in 1906 and a gas-powered model the next year.

These men and their companies reenacted the same familiar scenario—fierce competition for market share that eventually led to a merger. Their 1908 merger was actually Holt buying out the Best company.

Daniel Best's son, C. L. "Leo" Best, continued to work for the Holt company after the buyout; but after two years, in 1910, he began his own company, and the Best-Holt rivalry began all over again. The C. L. Best Gas Traction Company was founded at Elmhurst, California. In 1912, the new company offered its first crawler model as the Tracklayer.

The Holt versus Best contest was resolved yet again by Holt buying the new Best company in 1925 and folding it into the Caterpillar Tractor Company of Peoria, Illinois. From that time until today, it's been the home of the famous "Cat."

By 1940, the product line included motor graders, blade graders, elevating graders, terracers, and electrical generating sets. Heavy construction and industrial equipment became the focus of the company.

Engines built by Caterpillar have found their way into numerous OEM applications around the world. In 1953, a separate sales and marketing division was created for engine customers.

Caterpillar acquired United Kingdom–based Perkins Engines in 1997. This step, coupled with its existing engines operation, makes Caterpillar the world's leading diesel engine manufacturer.

Although Caterpillar tractors were originally designed primarily for agricultural applications and have always been used on fields and farms, it wasn't until the 1960s that the company once again targeted some of its crawlers for the agricultural market.

The biggest difference between the construction/industrial models and the agricultural models was the transmission speed. Beginning in the 1960s, the transmission was redesigned to provide speeds from 2.5 to 5 miles an hour for tillage drawbar work.

Three of the models in the 1960s that were typical of the agricultural models carried the designation "special application tractors." The D4D SA was produced from 1966 to 1968 and used a Caterpillar engine that was rated at 68-drawbar horsepower. Its serial number prefix was 20J, indicating that it was a special application, or agricultural vehicle.

A bit larger, the D5 SA ran from 1967 through 1977 and fell into the 90-drawbar horsepower class. The serial number prefix was 21J.

In 1970, the 125-drawbar horsepower D6C SA was introduced to the line. ROPS and comfort cabs were an option on these crawlers. There were also options in track gauge and track shoe widths.

Today, Caterpillar is again becoming a presence in the agricultural equipment field and promises to provide some of the most innovative future machines for agribusiness worldwide.

WAGNER

"From the heart of the nation's timber country" is the way Wagner Tractor's early advertisements proclaimed its roots in Portland, Oregon. The Wagner brothers started building complete mobile concrete mixing and elevating plants in 1940. Contractors were soon asking for additional construction equipment. This led to production of large four-wheel-drive tractors for the timber industry.

In the early 1950s, the company introduced big four-wheel-drive agricultural tractors that found a most receptive market in the large farming operations of the West. The peat, sand, mud, and hills of California, Arizona, and the Northwest were almost exclusively crawler territory until Wagner introduced a workable alternative. These tractors also found favor in large farming operations throughout the Plains states.

Mud and uneven terrain is no challenge for the Wagner four-wheel drive with Pow-R-Flex coupling and dual axle oscillation. This exclusive Wagner feature allowed all wheels to maintain positive down pressure. Even in short turns, full steering control with no loss of power was a feature noted in the Wagner literature. *Ervin Wagner, President of Wagner Tractor, Inc.*

151

Wagners were also used for industrial applications, as this 150-horsepower Wagner IND-14 demonstrates by pulling a Be-Ge WSS-85120 12-yard scraper. *Ervin Wagner, President of Wagner Tractor, Inc.*

The 1981 525/50 had a thirteen-forward-speed transmission, with partial range power shift. *Andrew Morland*

152

Waukesha engines were used in almost all Wagners before the 1956 production year. At that time a change was made to Cummins power.

Company literature, circa 1958–1959, listed these features:
- Four-wheel drive
- Four-wheel steering
- Planetary drive
- Both axles oscilate
- Center pin hydraulic steering with flow divider
- Wheels always track
- No power loss in turns
- Powered by Cummins Diesel engines
- Optional items include cabs, air conditioning, heater and defroster, and hydraulic tool-bar

The TR-6 Diesel used a Cummins diesel engine rated at 105 horsepower. Also produced in 1957 was the Wagner TR-9 Diesel. This unit was perhaps the first production articulated four-wheel-drive tractor made. The four-cylinder Cummins engine developed 87-drawbar horsepower. It had a 10-speed transmission and weighed 15,445 pounds.

In 1959, the Wagner TR-14A Diesel tested at 155-drawbar horsepower from a six-cylinder Cummins engine. It was fitted with a 10-speed transmission. This model was articulated, featured an enclosed cab, and weighed 21,225 pounds.

FWD Corporation of Clintonville, Wisconsin, purchased controlling interest in the Wagner Tractor Company in the early 1960s. Production continued at the Wagner facilities in Portland, Oregon.

Introduced in 1964, the FWD Wagner WA4 Diesel featured the GM two-cycle, four-cylinder diesel engine. It was turbocharged and rated at 97-drawbar horsepower. Its transmission was an eight-speed, and a cab was standard equipment. Other models between 1964 and 1969 included the WA-9, WA-14, WA-17, and WA-24.

In 1969, Deere and Company contracted for Wagner's entire production capabilities. The WA-14 and WA-17 were badged and painted appropriate for Deere and Company's line. The WA-14 was rated at 178-drawbar horsepower, and the WA-17 fell into the 220-drawbar horsepower area.

The contract into which Deere and Company and Wagner entered specified that if Deere and Company stopped buying tractors from Wagner, Wagner couldn't produce a competing four-wheel-drive tractor for five years. So when Deere and Company introduced its own four-wheel-drive unit in 1970, FWD Wagner was effectively out of the farm tractor business.

BIG BUD

Willie Hensler was a Wagner dealer in Havre, Montana, during the 1960s. When John Deere purchased Wagner's entire production during the late 1960s, Hensler found himself without a tractor to sell. So he began by repowering and rebuilding Wagner tractors to provide for his customers. This led to the design and production of the Big Bud tractor.

The engine, transmission, and radiator were built as an integral unit and mounted on skids that allowed the unit to be removed for servicing.

Built in 1981 by the Northern Manufacturing Company of Havre, Montana, the six-cylinder 1,150 ci turbocharged, intercooled Cummins diesel engine produces 525 horsepower. *Andrew Morland*

153

Big Bud's first tractor, the 1968 HN 250 diesel, used a six-cylinder Cummins engine that the manufacturer estimated at 310 horsepower. The articulated tractor weighed 34,000 pounds and was fitted with a 12-speed transmission.

The next Big Bud was the HN 320 Diesel introduced in 1970 and offered through 1978. It was advertised as producing 320 horsepower from a Cummins six-cylinder engine with turbocharger and intercooler. Transmission was a 12-speed with partial range Power-Shift. The tractor weighed in at 36,000 pounds. A cab was available.

Hensler sold the company in the mid-1970s. It was renamed the Northern Manufacturing Company.

The new owners continued expanding the tractor line and horsepower. In 1977, Big Bud rolled out the largest agricultural tractor ever built up to that time, the 780-horsepower 16V-747.

In 1982, the company filed for bankruptcy. Meissner Manufacturing Company purchased the assets of the company, and between 1982 and 1987 produced approximately 13 additional tractors.

It was evident that big high-horsepower tractors had become a vital part of the tractor industry in the 1960s and beyond. Major companies were either building their own or had bought out the competition. Big Bud couldn't compete, and production stopped on the Big Bud tractors in the 1980s.

M-R-S

Mississippi Road Supply began selling industrial tractors and heavy earth-moving equipment

This 525/50 is painted in the standard white color scheme. It is owned and used by farmer Richard Dahlgren on his large farm in western Minnesota at Bird Island. *Andrew Morland*

M-R-S has an industrial history that began in 1943, but it wasn't until the late 1950s that a strictly agricultural model was introduced. The M-R-S Model A-100 was introduced in 1964 as a high-horsepower, four-wheel-drive, four-wheel-steer agricultural tractor for large farming operations. Powered by a GM 6V-71N two-cycle valve-in-head diesel, it produced 215 engine horsepower. Photo courtesy Flora Public Library, Flora, Mississippi

155

The M-R-S A-100 was also available with a 552-ci Detroit diesel engine that was rated at 322 engine horsepower. This model was produced from 1964 to 1987 and came with a cab as standard equipment. Weighing in at 26,000 pounds, the M.L.S.R.P. was $146,195. Photo courtesy Flora Public Library, Flora, Mississippi

in 1943. At that time, the company was located in Jackson, Mississippi. As the demand for large four-wheel-drive agricultural tractors developed, the owners of Mississippi Road Supply began producing an all-wheel-steer agricultural unit in 1963.

The Model A-80 used a two-cycle, four-cylinder GM engine that produced 141 horsepower. The all-wheel drive was coupled to a 10-speed transmission.

J. W. Richardson, who worked at M-R-S as production manager from 1955 to 1968, recalled that at one time M-R-S built the largest rubber-tired tractor in the world. "When I left in 1968, we had a government contract to build equipment for the Marine Corps to use in Vietnam," Richardson said.

"We built more than 100 tractors and 100 scrapers that went behind these tractors to fulfill the contract. We built these in 'packages' that could be dropped by parachute from helicopters and then quickly reassembled on the ground.

"Our tractors had hydraulic couplers that enabled the snapping on of component systems quickly. The front axle would be in one package, the engine and dozer blade in another package, the rear axle in another package, and the rest in another package."

Richardson explained how the company was named: "The company was owned by L. R. Simmons. He owned a distributing company in Jackson, Mississippi, that sold Caterpillar and other heavy equipment brands. This company was named Mississippi Road Supply.

"Most people thought that the M-R-S name stood for Mississippi Road Supply, but Simmons had a daughter named Mariada Reynolds Simmons. It was my understanding that he really named the company after his daughter. M-R-S Manufacturing Company was completely different from the Mississippi Road Supply."

M-R-S fabricated all the levers, brackets, engine mounts, radiator mounts, and frames. In-house production also included hoods, dashes, fuel tanks, and fenders.

Axles were outsourced from Rockwell Standard Corporation. Most of the engines were from Cummins; however, a few International engines found their way into M-R-S tractors too. Radiators, hydraulic cylinders, seats, and steering wheels were also purchased from outside suppliers. Transmissions were outsourced from Rockwell, Fuller, and Spicer.

The A-80 was the first model. It was followed by the A-90, the A-100, A-105, and the A-110. The bigger the number the higher the horsepower.

"We also were making 30 to 40 agricultural tractors a year. We sent a large number of these units to U.S. Sugar in Miami, Florida. The agricultural tractors were red and those produced for the military were in military colors," Richardson said. "We built a 47-yard scraper too. The TVA, Tennessee Valley Authority, bought some of our tractors with these large scrapers behind them.

"But what kept the company going was military contracts, so when the war was over in Vietnam that put an end to military contracts.

When M-R-S took bankruptcy, its name, blueprints, and patterns were bought by Taylor Machine Works of Louisville, Mississippi.

A unit of impressive size, this M-R-S Industrial 250 came equipped with a GM 16V-71N 16-cylinder diesel engine that carried a factory-observed rating of 635 horsepower. The transmission was a power-shift with six forward and two reverse gears. M-R-S designed this vehicle to be used in strip mining operations for powering tandem bottom-dump wagons with capacities up to 140 net tons. *Photo courtesy Flora Public Library, Flora, Mississippi*

BIBLIOGRAPHY

American Society of Agricultural Engineers. *John Deere Tractors, 1918–1994.* St. Joseph, Michigan: American Society of Agricultural Engineers, 1994.

Baumheckel, Ralph, and Kent Borghoff. *International Harvester Farm Equipment: Product History 1831–1985.* St. Joseph, Michigan: American Society of Agricultural Engineers, 1997.

Broehl, Wayne G., Jr. *John Deere's Company.* New York: Doubleday and Company, 1984.

Erb, David, and Eldon Brumbaugh. *Full Steam Ahead: J. I. Case Tractors and Equipment 1842–1955.* St. Joseph, Michigan: American Society of Agricultural Engineers, 1993.

Fisher, Douglas Alan. *The Epic of Steel.* New York, Evanston, and London: Harper and Row, 1963.

Gibbard, Stuart. *The Ford Tractor Story: Dearborn to Dagenham 1917–1964.* Hutton, Driffield, United Kingdom: Joponica Press, 1998.

Halberstadt, April. *Case Photographic History.* Osceola, Wisconsin: MBI Publishing Company, 1995.

Halberstadt, April. *Oliver Tractor Photographic History.* Osceola, Wisconsin: MBI Publishing Company, 1997.

Herndon, Booton. *Ford: An Unconventional Biography of the Men and Their Times.* New York: Weybright and Talley, 1969.

Holmes, Michael S. *J. I. Case: The First 150 Years.* Milwaukee, Wisconsin: Case Corporation, 1992.

Huber, Donald S., and Ralph C. Hughes. *How Johnny Popper Replaced the Horse.* Moline, Illinois: Deere and Company, 1988.

Huxley, Bill. *Allis-Chalmers Agricultural Equipment.* London: Osprey Publishing, 1988.

Klancher, Lee. *Farmall Tractors.* Osceola, Wisconsin: MBI Publishing Company, 1995.

Klancher, Lee. *International Harvester Photographic History.* Osceola, Wisconsin: MBI Publishing Company, 1996.

Kudrle, Robert T. *Agricultural Tractors: A World Industry Study.* Cambridge, Massachusetts: Ballinger Publishing Company, 1975.

Larsen, Lester. *Farm Tractors 1950–1975.* St. Joseph, Michigan: American Society of Agricultural Engineers, 1981.

Leffingwell, Randy. *The American Farm Tractor.* Osceola, Wisconsin: MBI Publishing Company, 1991.

Leffingwell, Randy. *Classic Farm Tractors.* Osceola, Wisconsin: MBI Publishing Company, 1993.

Leffingwell, Randy. *Ford Farm Tractors.* Osceola, Wisconsin: MBI Publishing Company, 1998.

Leffingwell, Randy. *John Deere Farm Tractors.* Osceola, Wisconsin: MBI Publishing Company, 1993.

Macmillan, Don. *John Deere Tractors Worldwide.* St. Joseph, Michigan: American Society of Agricultural Engineers, 1994.

Macmillan, Don, and Roy Harrington. *John Deere Tractors and Equipment, Vol. 2.* St. Joseph, Michigan: American Society of Agricultural Engineers, 1991.

Macmillan, Don, and Russell Jones. *John Deere Tractors and Equipment, Vol. 1.* St. Joseph, Michigan: American Society of Agricultural Engineers, 1988.

Marsh, Barbara. *A Corporate Tragedy.* Garden City, New York: Doubleday and Company, 1985.

McCormick, Cyrus. *The Century of the Reaper.* Boston and New York: Houghton Mifflin Company, 1931.

Morland, Andrew. *Modern American Farm Tractors.* Osceola, Wisconsin: MBI Publishing Company, 1994.

Morland, Andrew. *Modern Farm Tractors.* Osceola, Wisconsin: MBI Publishing Company, 1997.

Morland, Andrew, and Peter Henshaw. *Allis-Chalmers Tractors.* Osceola, Wisconsin: MBI Publishing Company, 1997.

Morrell, T. Herbert, and Jeff Hackett. *Oliver Farm Tractors.* Osceola, Wisconsin: MBI Publishing Company, 1997.

Neufeld, E. P. *A Global Corporation.* Toronto, Canada: University of Toronto Press, 1969.

Nevins, Allan, and Ernest Hill. *Ford: Decline and Rebirth 1933–1962.* New York: Charles Scribner's Sons, 1962.

Nevins, Allan, and Ernest Hill. *Ford Expansion and Challenge 1915–1933.* New York: Charles Scribner's Sons, 1957.

Nevins, Allan, and Ernest Hill. *Ford: The Times, The Man, The Company.* New York: Charles Scribner's Sons, 1954.

Peterson, Chester Jr., and Rod Beemer. *Ford N Series Tractors.* Osceola, Wisconsin: MBI Publishing Company, 1997.

Peterson, Chester Jr., and Rod Beemer. *John Deere New Generation Tractors.* Osceola, Wisconsin: MBI Publishing Company, 1998.

Peterson, Walter F. *An Industrial Heritage: Allis-Chalmers Corporation.* Milwaukee, Wisconsin: Milwaukee County Historical Society, 1998.

Pripps, Robert N. and Andrew Morland. *Ford Tractors.* Osceola, Wisconsin: MBI Publishing Company, 1990.

Pripps, Robert N. and Andrew Morland. *Oliver Tractors.* Osceola, Wisconsin: MBI Publishing Company, 1994.

Sanders, Ralph W. *Vintage Farm Tractors.* Stillwater, Minnesota: Voyageur Press, 1996.

Sayers, Al. *Minneapolis-Moline Tractors.* Osceola, Wisconsin: MBI Publishing Company, 1996.

Stonehouse, Tom, and Eldon Brumbaugh. *J. I. Case: Agricultural and Construction Equipment 1956–1994.* St. Joseph, Michigan: American Society of Agricultural Engineers, 1996.

Swinford, Norm. *Allis-Chalmers Farm Equipment 1914–1985.* St. Joseph, Michigan: American Society of Agricultural Engineers, 1994.

Swinford, Norm. *A Guide to Allis-Chalmers Farm Tractors.* St. Joseph, Michigan: American Society of Agricultural Engineers, 1996.

Wendel, C. H. *The Allis-Chalmers Story.* Osceola, Wisconsin: MBI Publishing Company, 1993.

Wendel, C. H. *Encyclopedia of American Farm Tractors.* Osceola, Wisconsin: MBI Publishing Company, 1992.

Wendel, C. H. *Oliver: Hart-Parr.* Osceola, Wisconsin: MBI Publishing Company, 1993.

Wendel, C. H. *150 Years of International Harvester.* Osceola, Wisconsin: MBI Publishing Company, 1993.

Wendel, C. H. *150 Years of J. I. Case.* Osceola, Wisconsin: MBI Publishing Company, 1994.

Wendel, C. H. and Andrew Morland. *Massey Tractors.* Osceola, Wisconsin: MBI Publishing Company, 1992.

Wendel, C. H. and Andrew Morland. *Minneapolis-Moline Tractors 1870–1969,* Osceola, Wisconsin: MBI Publishing Company, 1990.

Williams, Michael. *Ford and Fordson Tractors.* London, England: Blandford Press, 1985.

INDEX

A Corporate Tragedy, 52
Allis-Chalmers, 8, 9, 73–87
   100 Series, 83, 84, 86
   12-20, 76
   15-25, 76
   15-30, 76
   6-12, 76, 83
   7000 Series, 86
   C, 78
   CA, 78
   D Series, 82
   D10, 77
   D12, 79, 86
   D14, 78, 79
   D15, 76
   D17, 78, 79
   D17D, 78
   D19, 80
   D21, 82–84
   E18-30, 76
   E20-35, 76
   E25-40, 76
   ED40, 85
   G, 78
   Hi-Crop, 85, 86
   Model 10-18, 76
   One-Ninety XT, 84, 85
   One-Sixty, 85
   Power-Director, 79, 80, 82, 84
   Power-Shift, 78
   Roll-Shift, 78
   Series II, 77, 78, 83
   Series III, 85, 86
   Series IV, 73
   Snap-Coupler, 84
   Sugar Babe, 82, 83
   Two-Twenty, 85
   U, 76
   W, 76
   WC, 76
   WD, 78
   WD45, 78
   WF, 78
Adams Wagon Company, 141
Adams, Wilbur Henry, 122
Advance-Rumley Thresher Company, 76
AGCO Allis, 87
Agricultural Tractors: A World Industry Study, 11
Allied Products Corporation, 145
Allis, Edward P., 73
Allis-Gleaner Corporation (AGCO), 87, 102, 145
American Seeding Machine Company, 121
American Tobacco Company, 41
American Tractor Company (ATC), 60
Anderson, Jo, 41
B.F. Avery, 136
   Model V, 136
Bain Company, 89
Baker Company, 84
Baumheckel, Ralph, 44, 46, 51
Bay State Iron Works, 73
Be-Ge, 122
Best Manufacturing Company, 150
Best, C. L. "Leo", 150
Best, Daniel, 150
Best, Tracklayer, 151
Big Bud, 153, 154
   16V-747, 154
   525/50, 152–154

HN 250, 154
HN 320, 154
Birdsell Manufacturing Company, 76
Black, Sir John, 90
Brantford Carriage Company, 141
Brantford Plow Works, 141
Brock, Harold, 26, 36, 114
Brumbaugh, Eldon M., 65
Buda Company, 84
Bull Tractor Company, 89, 133
Bullock Electric and Manufacturing Company, 75
C. L. Best Gas Traction Company, 151
Call, Delmar W., 75
Case Credit Corporation, 60
Case IH, 55
Case Threshing Machine Company, 65
Case, 9, 10, 13
   1030 Wheatland Western Special, 9
   1030, 65, 69
   1200, 71
   130, 9, 60
   180, 9
   30 Series, 60
   400 Series, 60
   430 LCK, 9
   430, 63
   530, 9, 64
   630, 64
   630D, 60
   730, 9, 64, 65
   800, 59
   830 Hi-Crop, 9
   830, 62
   930 LP, 9
   930, 64, 65
   Agri-King, 69
   Alphabet Line, 57
   B Series, 58
   Case-O-Matic, 57–60, 62, 64, 65, 70
   Comfort King Series, 64, 65, 67, 68
   Draft-O-Matic, 65
   Dual-Range, 64
   Eagle Hitch, 57
   Hi-Crop, 57, 62
   Orchard and Grove, 9
   Orchard, 9
   Power-Shift, 66
   Row-Crop, 57, 58, 61, 62, 67–69
   Snap-Lock Eagle Hitch, 64
   Tripl-Range, 58, 60, 61, 63, 64
   World Premiere, 70
Case, Jackson I., 60, 65
Case, Jerome Increase, 57, 60, 71
Caterpillar, 149–151
   D2, 150
   D4D SA, 151
   D5 SA, 151
   D6, 150
   D6C SA, 151
Chalmers, William J., 73
Charter, 7
Chris-Craft Outboard Motors, 122
Chrysler, 90
Clark, 147
Clausen, Leon R., 10
Cleveland Tractor Company, 122
Cockshutt Plow Company, 141
Cockshutt, 124, 130, 141–145
   30, 142
   40, 142

500 Series, 142, 143
540 Wide Adjustable, 143
550 Standard, 143
560 Standard Diesel, 144
570 Standard Diesel, 144
570 Standard Gasoline, 144
570 Super Diesel, 144
770, 144
880, 144
D-50, 142
E2, 142
E5, 142
Cockshutt, James G., 141
Colt Manufacturing Company, 62, 66
Corbin Disc Harrow, 89
David Brown, 90, 125, 126
Decker and Seville, 81
Decker, Charles, 73
Deere & Company, 9, 10, 13–39
Deere, Charles, 13
Deere, John, 13, 39, 57
Deering Harvesting Company, 41
Deutz-Allis, 87
Deyo-Macey Company, 89
Diesel, Rudolf, 43
Dreyfuss, Henry, 30
Droegemueller, Wally, 122
Dushane, Wally, 15
Elwood Engineering Company, 106
F. Perkins Ltd., 92–94
Falk Company, 76
Falk, Herman, 75
Falk, Otto H., 75
Farquhar-Iron Age, 122
Ferguson, Harry, 10, 90, 103, 107, 138
Ferguson, TE-20, 92
Ferguson-Brown, 90
Fiat, 125–127
Fiat-Allis, 87
FiatGeotech, 119
Firestone, 80
Fletcher, Ed, 16
Ford, 8, 105–119, 134, 149
   1801 Series, 118
   2000 Series, 116, 118
   3000, 116, 117
   3400, 118
   3500, 118
   4000 Series, 116, 118
   4400, 118
   4500, 118
   5000, 116
   541, 108
   600 Series, 112
   601, 114
   800 Series, 112
   8N, 110
   Commander 6000, 115
   E27N, 110
   NAA Golden Jubliee, 112
   Power-Shift, 109
   Powermaster, 112, 118
   Select-O-Speed, 107, 109, 111, 114–117
   Workmaster, 111, 112
   World Tractor, 115, 117
Ford, Henry, 9, 10, 90, 105, 106, 108, 119
Ford, Henry, II, 92, 112
Ford-Ferguson, 13, 90
Ford-New Holland, 119
Fordson, 9

All-Around, 106
Model F, 105
Model N, 106
N Series, 108
Frank G. Hough Company, 48
Froelich, 7
Frost and Wood, 141
Fuller, 157
Funk Reversomatic, 126
FWD Corporation, 153
FWD Wagner
   WA-14, 153
   WA-17, 153
   WA-24, 153
   WA-9, 153
   WA4 Diesel, 153
General Motors, 9, 98
Gentner, Del, 95, 99, 103
Gleaner Harvester Corp., 84, 87
Gleeson, Danny, 16
Grand Detour Plow Company, 65
Great Depression, 9, 10, 43
Grede, William, 61, 65
Green, Jerome K., 71
Hansen, Merlin, 15
Harris, Alanson, 89, 103
Hart, Charles, 121
Hart-Parr, 121, 141
   17-30, 121
   22-40, 121
   No. 1, 121
   No. 2, 121
Hennessy, Patrick, 110
Henry Ford and Son Inc., 405
Henry Mfg. Co., 84
Hensler, Willie, 153
Hess, Chris, 16
Hesston, 71, 87
Hewitt, Bill, 16, 26, 29, 36
Hoar Shovel Company, 76
Holmes, Michael, 69
Holt Manufacturing Company, 150, 151
Holt, Benjamin, 150
Holt, Charles, 150
Horner, Don, 114, 115, 117
Hough Manufacturing, 45
Hussey, Obed, 41, 44
Hy-Way Machine and Manufacturing Company, 76
Hydraulic Engineering Company, 149
Industrial Dufermex, S.A., 84
Ingersol Tractors, 62
International Harvester Company (IHC), 7, 9, 10, 13, 41–55, 71, 132,
   149, 157
   100 Series, 44
   1206, 48
   130, 48
   240, 48
   350, 48
   404, 46, 48
   4100, 46, 49
   4300, 48, 49
   460, 44
   504, 48
   560, 45
   706, 49, 51
   806, 49, 61
   A Series, 44
   B Series, 44
   Category II, 55

F-12, 43
F-14, 43
F-20, 43
F-30, 43
Farm Equipment Engineering and Research Center (FEREC), 44
Farmall 1206 Turbo Diesel, 51
Farmall Cub, 44
Fast-Hitch, 48
H Series, 44
Hi-Crop, 42
Hundred Series, 46, 48
M Series, 44
Mogul 8-16, 42
Power-Shift, 51
Titan 12-25, 42
Touch Control, 48
Type A, 41
W-12, 43
W-30, 43
W-40, 43
WD-40, 43
International Harvester Tenneco, 149
J.I. Case Company, 55, 57–71
J.I. Case Plow Works, 65, 89
J.I. Case Threshing Machine, 133
J.I. Case: The First 150 Years, 69
Jeep, 134
John Deere
   1010 Crawler, 21
   1010, 16, 23
   20 Series, 13
   2010, 16, 23
   2510, 23
   2520, 24
   30 Series, 13, 38
   3010, 16, 20, 23
   3020, 20, 23
   320, 13
   40, 13
   4000, 24
   4010, 16, 20, 23
   4020, 20, 23, 31
   40C, 13
   420, 13
   50, 13
   5010, 23
   5020, 23
   520, 13
   60, 13
   6030, 38
   620, 13
   70, 13
   7020, 36
   720, 13
   7520, 36
   80 Diesel, 13
   820, 13, 23
   Generation II, 38
   Hi-Crop, 24, 25, 33
   I Models, 20
   Model A, 13
   Model C, 13
   Model D, 9, 13, 20
   Model M, 13
   Model R, 13
   New Generation, 14–16, 18, 20, 23–26, 30–32, 36, 38
   Power-Shift, 23–25, 28, 31–33, 35
   Roll-Gard, 24, 35
   Roll-O-Matic, 17, 22, 27

Row-Crop Utility, 33
Row-Crop, 16, 17, 22, 23, 25, 27, 29, 33, 35
Sound-Gard, 29
Syncro-Range, 23–25, 29, 33, 35
Johnston Harvester Company, 89
Joy-Wilson Company, 133
Kemp Manure Spreader Company, 89
Kern County Land Company, 63, 66, 71
Ketelsen, James, 61, 70, 71
Klockner-Humboldt-Deutz AG, 87
Kudrle, Robert T., 11
LaCrosse Plow Company, 76
LaPlant-Choate Mfg. Co., 84
Lindeman Manufacturing Company, 20
Loewy, Raymond, 43, 142
Mack, Mike, 15, 26
Marsh, Barbara, 52
Massey Manufacturing Company, 89
Massey, Daniel, 89, 103
Massey-Ferguson, 9, 87, 89–103
   100 Series, 99
   65, 95
   Diesel-Matic, 98
   MF-1080, 101
   MF-1100, 97, 100
   MF-1130, 97, 101
   MF-130, 99, 100
   MF-135, 97, 98, 99, 103
   MF-148, 100
   MF-150, 97, 99, 100
   MF-165, 97, 98, 100
   MF-175, 97, 100
   MF-180, 100, 102
   MF-25, 95
   MF-35, 99
   MF-50, 97
   MF-65, 97, 98
   MF-85, 95, 98, 99
   MF-88, 98, 100, 101
   MF-95, 98
   MF-97, 98
   MF-98, 98
   Multi-Power, 98
   Show of Progress, 99
   Super 90, 98
Massey-Harris, 9, 22, 44, 65, 89
   22 Row-Crop, 92
   25-40, 89
   30 Pony, 92
   Big Bull, 89
   Challenger, 89
   General Purpose four-wheel-drive tractor, 89
   Harvest brigade, 92
   Hydraulic Depth-O-Matic Control, 92
   Model 101, 90
   Pacemaker, 89
Massey-Harris-Ferguson Ltd., 92
McCormick Harvesting Machine Company, 41
McCormick International
   B-275, 48
   B-414, 48
McCormick, Cyrus Hall, 41, 44, 57
McCormick, Cyrus, Jr, 41
McCormick, Leander, 41
McCormick, Nancy "Nettie", 41
McCormick, William, 41
McCormick-Deering, 44
McRae, Ivan, 145
Meissner Manufacturing Company, 154

159

Micromatic Hone, 84
Milwaukee Harvester Company, 41
Minneapolis Steel and Machinery Company, 130, 133
    Big Bull, 133
    Little Bull, 133
    Twin City 12-20, 134
    Twin City 16-30, 134
    Twin City, FTA, 134
    Twin City, KTA, 134
    Twin City, MTA, 134
    Twin City Model 40, 133
Minneapolis Threshing Machine Company, 130
    15-30, 134
    17-28, 134
    20-40, 134
    25-50, 134
    27-44, 134
    35-70, 134
    40-80, 134
    Universal 20-40, 134
Minneapolis-Moline, 9, 10, 98, 130–141, 145
    335, 136
    4 Star Super, 10
    445, 136
    4T-1600, 10
    5-Star, 136
    A4T-1600, 10, 140
    G, 136
    G1000 Vista, 10, 140
    G1000, 139
    G1050, 10
    G1350, 140
    G550, 130
    G705, 139
    G706, 10, 139
    G707, 139
    G708, 139
    G940, 130
    G950, 140
    GT, 136
    GTA, 136
    GTB, 136
    GTC, 136
    Heritage Model, 138
    Jet Star 3 Super, 10
    M-5 Series, 139
    M-670, 139
    M5, 10
    M602 Propane, 10
    M670, 10
    Motrac, 10
    Powerline Series, 136
    Sport Open Model Roadster, 135
    Super M-670, 139
    U Series, 135, 136
    U-302, 139
    U302 Super, 10
    UDLX Comfortractor, 135
    Uni-Baler, 138
    Uni-Foragor, 138
    Uni-Harvester, 138
    Uni-Picker, 138
    Uni-Tractor, 138
    Visionline, 135
    White G-950, 10
Mississippi Road Supply (M-R-S), 154–157
    A-100, 155–157
    A-105, 157
    A-110, 157
    A-80, 156, 157
    A-90, 157
    Industrial 250, 157
Moline Implement Company, 132
Moline Plow Company, 130, 134
Moline Universal, 132
Monarch Tractor Company, 76

Morrell, T. Herbert, 121
New Idea, 71
Niagara Falls Power Company, 81
Nichols and Shepard Threshing Machine Company, 121
Nordyke and Marmon Company, 76
Northern Manufacturing Company, 154
Oldfield, Barney, 80
Oliver Chilled Plow Works, 121
Oliver Farm Equipment Company, 121
Oliver Farm Tractors, 121
Oliver, 9, 11, 98, 121–130, 141, 145
    1250, 126, 127
    1255, 127
    1355, 127
    1450, 127
    1550, 129
    1555, 130
    1600, 123, 126
    1650, 125, 127
    1655, 130
    1750, 123, 127
    1800, 123, 124
    1800A, 125
    1800B, 125
    1800C, 125
    1850, 127, 129
    1855, 130
    1900, 123, 124, 127
    1900A, 15
    1900B, 125
    1900C, 125
    1950T, 129
    2050, 129
    2150, 129
    440, 125
    500, 125
    600, 126
    Cletrac, 124
    Fleetline, 122, 126
    Hi-Crop, 125
    Hydra-Power Drive, 125–127, 130
    Hydraul-Shift, 126, 129, 130
    Power Booster Drive, 126
    Power-Shift, 125
    Riceland, 125, 130
    Ridemaster, 122
    Row-Crop, 125
    Super 66, 122
    Super 77, 122
    Super 88, 122
    Wheatland, 125, 126, 130
Oliver, James D., 121
Oliver, Joseph D., 121
Olsen, Sid, 16
Pakosh, Peter, 149
Parr, Charles, 121
Parrett Tractor Company, 89
Patterson Wisner Company, 89
Perkins Engines, 151
Perry, Sir Percival, 105
Pinardi, Edward, 114, 117, 118
Pittsburgh Transformer Company, 76
Plano Harvesting Company, 41
Red Foley group, 99
Reeves and Company, 133
Retterath, Barney, 125, 127, 129
Reynolds, Edwin, 73
Rice, Victor A., 103
Richardson, Jw, 156, 157
Rockwell Standard Corporation, 157
Rojtman, Marc, 60, 61, 70
Rollover Protective Structures (ROPS), 11, 26
Ronning, Martin, 138
Rugen, Vern, 16
S. Morgan Smith, 84
Scholl, Willis, 80

Schwager-Wood Company, 84
Scott, David, 86
Seville, James, 73
Sherman and Shepard Company, 106
Siemens-Allis, 87
Simplicity Mfg. Co., 84
South Bend Iron Works, 121
Sperry New Holland, 119
Spicer, 147, 157
Stearns Motor Company, 76
Steiger Tractor Company, 71, 119, 147–149
    1 FWD, 147
    1200, 147
    1250, 148
    1700, 147
    2200, 147, 148
    3300, 147, 148
    850, 148
    Wildcat, 148
Steiger, Douglas, 147
Steiger, Maurice, 147
Stevens, Brooks, 78
Stevenson, Bob, 80
Swinford, Norm, 80, 81, 86
T.C. Pollard Pty. Ltd., 84
Taylor Machine Works, 157
Tenneco Inc., 55, 69–71
The Furrow, 24
Tooling and Production, 70
Tractomotive Corp., 84
Travell, Dr. Janet, 30
U.S. Steel, 41
United Tractors and Farm Equipment, 76
Universal Tractor Company, 132
    D, 132
    JT, 134
    JTO Orchard, 134
    Z, 134
    ZN, 134
Valley Iron Works Corp., 84
Vaughan, Lee, 137–139
Verity Plow Company, 89
Versatile
    125 4WD, 149
    145 4WD, 149
    D100, 149
    D118, 149
    G100, 149
Vietnam War, 10
Wagner, 151–153
    IND-14, 152
    TR-14A Diesel, 153
    TR-6 Diesel, 153
    TR-9 Diesel, 153
Wallis, 89
Warder, Bushnell, and Glessner Company, 41
Waterloo Boy, 9, 13, 20
Weber, Vincent P., 124
Wheeler, Warren, 141, 144
White, 11, 87, 121–145
White-New Idea Farm Equipment Company, 145
Wielage, Don, 15
Willys, John N., 134
Willys-Overland Motors, 134
Wiman, Charles, 26
World War I, 8, 105
World War II, 9, 10, 46, 66, 92
Worthington Pump and Machinery Corporation, 76

Model and Owner Index (Make, Year, Model, Owner, Page Number)

Allis-Chalmers
1960, D12, Edwin Karg, 74
1961, D12, Edwin Karg, 77
1961, D15, Edwin Karg, 75
1963, D12 Series II, Edwin Karg, 78
1963, D19, Edwin Karg, 79
1964, D21, Edwin Karg, 8, 81
1964, I40 Industrial, Edwin Karg, 87
1965, D12 Series 3, Edwin Karg, 85
1966, D17 Series 4, Edwin Karg, 72

Case
1960, 200, J. R. & Jay Gyger, 58
1962, 430 LCK, J. R. & Jay Gyger, 61
1962, 630 Orchard, J. R. & Jay Gyger, 59
1963, 730, J. R. & Jay Gyger, 61
1964, 830 Comfort King, J. R. & Jay Gyger, 56
1965, 530 Utility, J. R. & Jay Gyger, 63
1966, 1200, J. R. & Jay Gyger, 63
1967, 730 Orchard and Grove, J. R. & Jay Gyger, 64
1967, 930 Comfort King, J. R. & Jay Gyger, 67
1967, 930 Western Special Wheatland, J. R. & Jay Gyger, 66
1968, 1030 Comfort King Western Special, J. R. & Jay Gyger, 68
1969, 770 Agri King, J. R. & Jay Gyger, 69
Various garden tractors, J. R. & Jay Gyger, 62
Various models, J. R. & Jay Gyger, 9

Cockshutt
1960, 540, Chris Tepoel, 141
1960, 550, Darius Harms, 143
1960, 560, Don Lamb, 142
1960, 570 Wheatland, Don Lamb, 144
1961, 550, Chris Tepoel, 145
1961, 550, Darius Harms, 144
1961, 570, Don Lamb, 143
1968, 1650, Roger & Gene Uhglenhake, 145

Ford
1959, 841 Powermaster, Roger W. Elwood, 106
1959, 871 Powermaster, Roger W. Elwood, 115
1960, 971 LP, Dwight Emstrom, 109
1961, 6000 Commander LP, Dwight Emstrom, 111
1961, 671 (2000), Roger W. Elwood, 107
1962, 961, Paul Martin, 108
1964, 2000, McGinn Sales, 110
1964, 4000, Paul Martin, 104
1964, Military Tug, Ron Stauffer, 113
1967, 6000 LP, Dwight Emstrom, 8

International-Harvester
1960, 340 Hi-Crop, Jerry Mez, 42
1960, T-340 Crawler, Jerry Mez, 43
1961, 4300, Jerry Mez, 45
1961, 660 Wheatland, Darius Harms, 43
1964, 2806 Industrial, Jerry Mez, 46
1966, 1206 Wheatland, Jerry Mez, 47
1966, 1206, Jerry Mez, 47
1966, 806, Jerry Mez, 40
1966, TB25 Series B Crawler, Darius Harms, 49
1967, 806, Jerry Mez, 50
1968, 1256, Jerry Mez, 52
1968, 2856 Industrial, Jerry Mez, 53
1968, 856, Jerry Mez, 54
1968, 856, Jerry Mez, 7
1969, 756, Jerry Mez, 54
1969, 826, Roger & Gene Uhglenhake, 55

John Deere
1960, 1010 Utility, Wayne Findley, 14
1960, 3010, 4010, Kenny Smith, 6
1960, 3010, Kenny Smith, 27
1960, 4010, Darold Sindt, 30
1960, 4010, Kenny Smith, 29
1960, 8010, Walter & Bruce Keller, 39
1960, 8010/8020, Mike & Rick Hoffman, 38
1960, 830 Rice Special, Dan & Ken Peterman, 14
1961, 1010, Wayne Findley, 15
1961, 1010, Wayne Findley, 17
1961, 2010 Hi-Crop, Randy Addington, 12
1961, 2010 Hi-Crop, Randy Addington, 24
1961, 4010 Wheatland, Jon Kinzenbaw, 30
1962, 4010, Jon Kinzenbaw, 29
1963, 1010 Industrial, Dwight Emstrom, 19
1963, 1010 Industrial, Glen Knudson, 18
1963, 5010, Darold Sindt, 37
1964, 1010 Crawler, Randy Griffin, 20
1964, 4020, Jack Purinton, 32
1965, 2010, Mel Kopf, 22
1965, 3020 Wheatland, Larry Maasdam, 28
1965, 4020 Hi-Crop, Glen Knudson, 33
1965, 4020, Randy Addington, 33
1966, 2510 Hi-Crop, Jeff McManus, 25
1966, 2510, McGinn Sales, 25
1966, 3020, Jon Kinzenbaw, 28
1966, 4020 LP, Darold Sindt, 32
1966, 5020, Jon Kinzenbaw, 38
1969, 2520, Jack Purinton, 27
1969, 4000 Low Profile, Frank Rochowiak, 35
1969, 4000, Brian H. Thompson, 35
1969, 4000, Jon Kinzenbaw, 35
1969, 4000, Kenny Smith, 35
1969, 4020, Brian H. Thompson, 34

Massey-Ferguson
1960, 88 LP, Lawrence Meyers, 90
1960, 98, Ken & Kent Peterman, 91
1961, 202 Industrial, Keith Oltrogge, 93
1961, 95 Super, Dan & Ken Peterman, 9
1961, 95 Super, Ken & Kent Peterman, 88
1962, 35 Utility, McGinn Sales, 93
1962, 65, McGinn Sales, 94
1963, 25, Keith Oltrogge, 99
1963, 97, Jim Schlappi, 96
1964, 135, Warren Randall, 103

Minneapolis-Moline
1960, 4 Star Super, Roger, Eugene, Martin, and Gaylen Mohr, 133
1961, M5, Roger, Eugene, Martin, and Gaylen Mohr, 134
1964, G706, Roger, Eugene, Martin, and Gaylen Mohr, 136
1964, M602, Roger, Eugene, Martin, and Gaylen Mohr, 135
1964, M670 Super, Roger, Eugene, Martin, and Gaylen Mohr, 135
1965, U302, Roger, Eugene, Martin, and Gaylen Mohr, 136
1966, Jet Star III, Roger, Eugene, Martin, and Gaylen Mohr, 137
1969, A4T-1600 Plainsman Heritage, Roger, Eugene, Martin, and Gaylen Mohr, 139
1969, G1000 Vista, Roger, Eugene, Martin, and Gaylen Mohr, 140
1969, G1050, Roger, Eugene, Martin, and Gaylen Mohr, 137
1969, G950, Roger, Eugene, Martin, and Gaylen Mohr, 138
1969, Heritage A4T-1600, Roger, Eugene, Martin, and Gaylen Mohr, 138
Various models, Roger, Eugene, Martin, and Gaylen Mohr, 10

Oliver
1960, 1900 Wheatland Special, Lyle Dumont, 126
1960, 990, Chris Tepoel, 122
1961, 880, Roger & Gene Uhglenhake, 123
1964, 1600, Duane Peterson, 11
1964, 1600, Lyle Dumont, 124
1965, 1750, Roger & Gene Uhglenhake, 127
1965, 1950, Roger & Gene Uhglenhake, 129
1967, 1850, Roger & Gene Uhglenhake, 128
1968, 1650, Lyle Dumont, 120
1968, 2150, Roger & Gene Uhglenhake, 131
1969, 1655, Roger & Gene Uhglenhake, 132
1969, 1755, Roger & Gene Uhglenhake, 131
1969, 1850, Roger & Gene Uhglenhake, 128
1969, 1955, Roger & Gene Uhglenhake, 132
1969, 2050, Roger & Gene Uhglenhake, 130

Steiger
1967, 2300, Darius Harms, 148

Versatile
1967, 118, Darius Harms, 146